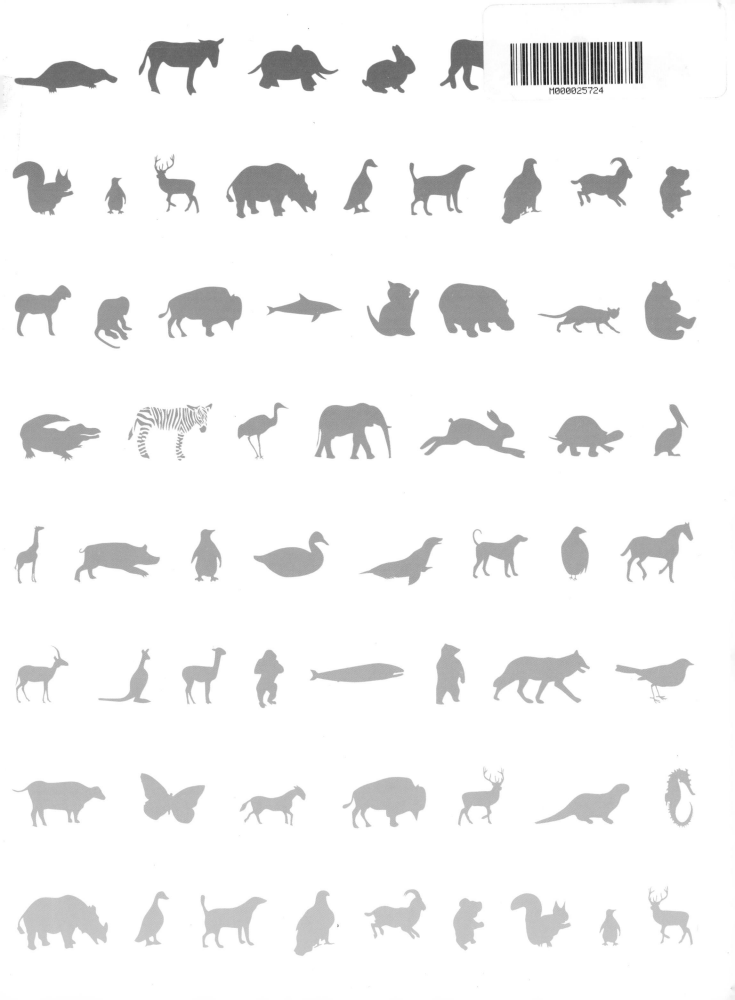

À la recontre des animaux
Translated by Gina Ivasuc

Copyright © EDITIONS CARAMEL S.A.
Otto de Mentockplein 19
1853 Strombeek-Bever, Belgium

036

ISBN 1-59496-035-6

Library of Congress Cataloging-in-Publication Data

Warnau, Geneviève, 1962-
  [À la rencontre des animaux. English]
  The encyclopedia of animals / by Geneviève Warnau.
      p. cm.
  Includes index.
  ISBN 1-59496-035-6 (hardcover : alk. paper)
  1. Animals--Encyclopedias, Juvenile. I. Title.
  QL49.W37 2005
  590'.3--dc22
                          2005000673

Printed in Slovakia

10 9 8 7 6 5 4 3 2 1

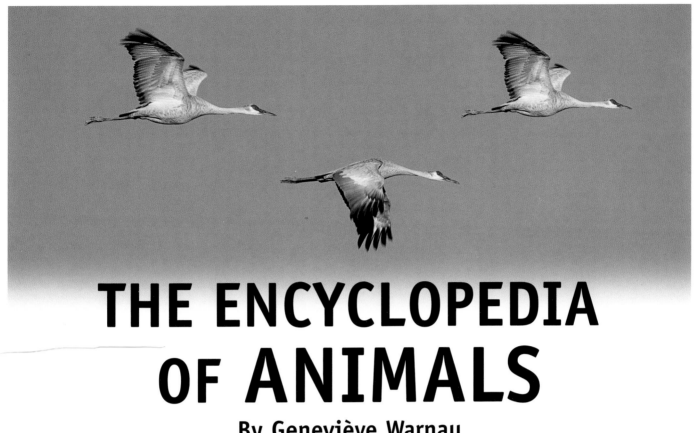

THE ENCYCLOPEDIA OF ANIMALS

By Geneviève Warnau

Published by Teora USA, Chevy Chase, Maryland, USA

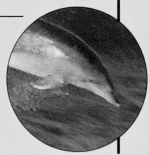

This encyclopedia classifies animals according to their habitat. You must consider, however, that some of them can live in many habitats. As far as possible, you will find such information in the text dedicated to each animal.

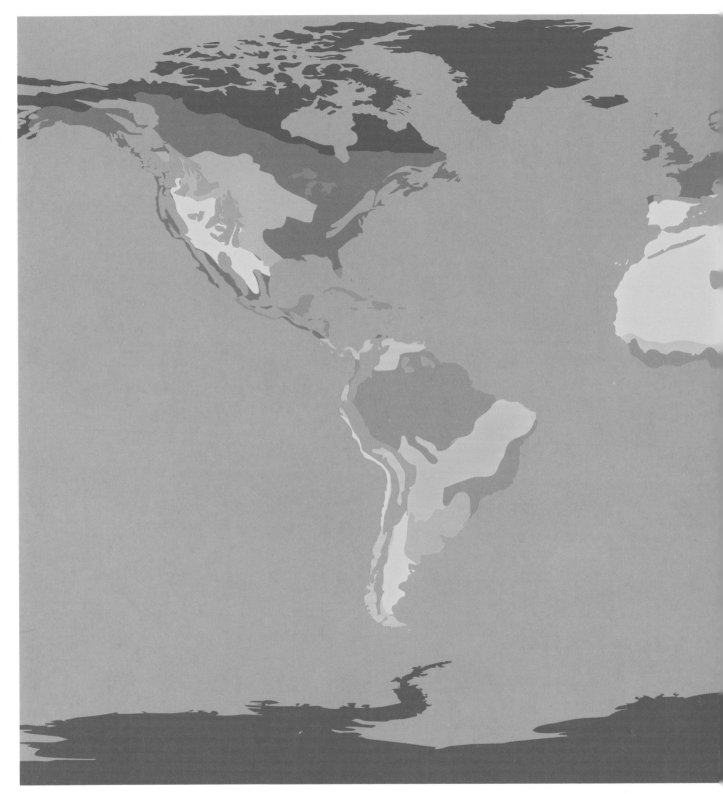

A Map of the Main Natural Environments

Seas and Oceans

Polar Regions

Mountains

Woods and Forests
(i.e., the cold forests, the woodlands, the temperate forests)

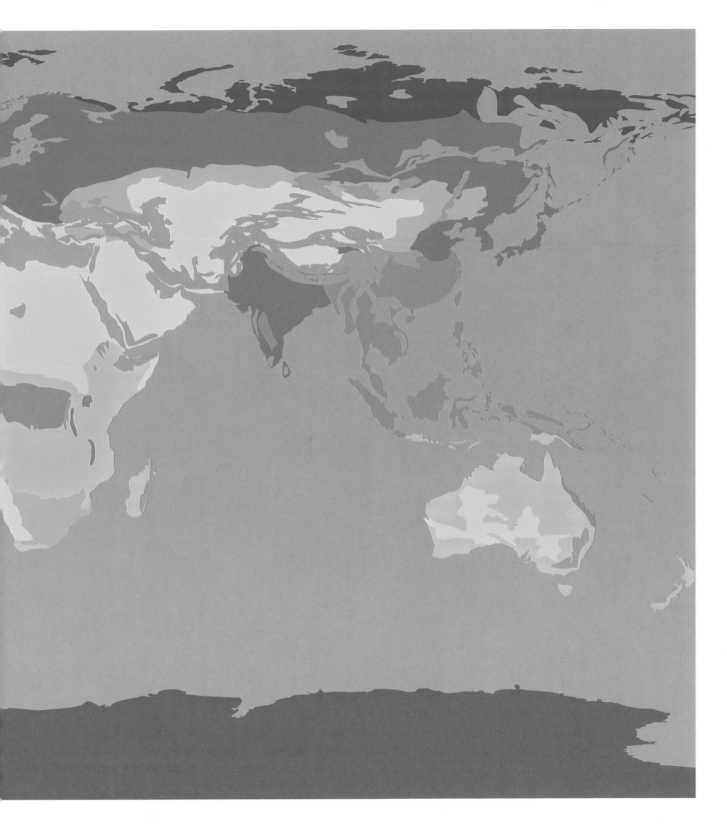

Deserts
(i.e., desert and
semidesert regions)

Savanna
(i.e., the savanna and the tropical
steppes)

Lakes

Rainforests

Grassy Plains and
Prairies

In the Savanna

THE LION

The lion is the only member of the cat family that lives and hunts in a group. The male, often called "king of the beasts" because of his beautiful mane, reigns majestically over the female companions of his pride. To keep intruders away from his territory, the lion lets out dreadful roars that can be heard more than six miles away. Nevertheless, he is extremely lazy, spending almost 20 hours each day sleeping or resting. The pride is actually headed by the lioness. She also hunts and rears the cubs. The lion cubs are like little fur balls and spend the first three months of their life with their mother, apart from the rest of the pride. At three years of age, the young adult males are chased away by the dominant male. They then try to displace the dominant male of another pride.

THE ELEPHANT

There are two species of elephants: the Asian elephant and the African elephant. The latter is pictured in these photos. Did you know that the elephant is the largest land animal alive nowadays? At birth, the little elephant already weighs more than 220 pounds. When it grows up, it will weigh up to 13,000 pounds and will stand about 13 feet tall. The elephant is so heavy that it must sleep standing up on all fours and cannot jump over even the smallest obstacle. The elephant is a peaceful animal. If threatened, however, it does not hesitate to charge its aggressor at full speed, spreading its huge ears, which makes it very impressive and sometimes very dangerous. Its ears are very special, because they help it sweat. The elephant drinks and eats with its trunk, which has two very nimble "fingers" at the end. Sometimes it uses its tusks to find food. The elephant has ivory tusks; they are actually very well developed teeth (incisors).

What a splendid animal is the cheetah! It is the aristocrat of the savanna. Its body is specially designed for running. It is actually the fastest terrestrial animal in the world. A cheetah can reach the incredible running speed of 68 miles per hour, and this is why it can hunt very fast prey such as the gazelle. And yet, despite its record speed, the cheetah misses nine times in 10 when hunting, just like most predators. Unlike other cats, the cheetah hunts in the daytime, and its claws are not completely retractile; they allow it to cling to the ground and to accelerate rapidly or to turn around suddenly in the full course of running. Its long tail is very useful in keeping its balance during these wild races. When it does not hunt, the cheetah spends its time in its lair; the female gives birth to four to six cubs, each weighing no more than an orange. At the age of one and a half, the cubs leave their mother, but they stay together until they become good hunters themselves.

THE CHEETAH

THE GIRAFFE

The giraffe is the tallest animal on the planet. The giraffe is 19 feet tall and dominates the savanna from this height. It watches over its surroundings, detecting the approach of great predators immediately. Because of its considerable height, the giraffe can find its favorite food, namely thorny acacia leaves, which it grabs with its long, hard tongue high up in the trees. Giraffes get the water they need from these juicy leaves and can sometimes live for a long time without drinking water. The giraffe sleeps standing up on all fours, but not more than 20 minutes a day, and then only for short periods of three or four minutes. At birth, the "little" giraffe calf is already 6.5 feet tall. The calf is welcomed from the very beginning into the herd, which can number as many as 40 individuals. The giraffe has long been considered a mute animal, but today we know that it can produce slight grunts.

THE OSTRICH

tanding about 10 feet tall and weighing about 300 pounds, the ostrich is the largest and the heaviest bird in the world. It cannot fly, but its powerful legs enable it to run as fast as a galloping horse. It can even carry a man on its back! The ostrich lives in flocks formed of one male and three or four females. During the mating period, the male digs a hole one meter in diameter in the ground. This hole acts as a kind of nest, where all the females lay about seven or eight huge eggs weighing three pounds each (the equivalent of about 25 hen's eggs!). The male and the dominant female of the group take turns incubating the eggs for 42 days. After the chicks hatch, the entire flock feeds them. The ostrich feeds on roots, leaves, insects, lizards, and small rodents, which it swallows without chewing. Presently, more and more ostriches are raised on farms, because their meat is considered highly desirable.

THE ZEBRA

Man could never domesticate this little striped horse! Zebras live in the wild, in families called "herds," led by a stallion. The various families share a vast territory and are distinguished by their stripe pattern, sounds, and smell. Each individual has distinctive stripes by which zebras identify themselves. You can compare these stripes to a sort of identity card or a giant "bar code." In addition, the stripes act as a protective camouflage against predators (especially lions). The zebra most often feeds on grass and sometimes on leaves or bark, and it must drink two to 2.6 gallons of water every day. The female gives birth to a foal after a one-year gestation period. The protective father never hesitates to defend his young when they are menaced by any wild animal. He fights by leaping fiercely and flashing his hooves, making even the lionesses stay away.

THE HIPPOPOTAMUS

The hippopotamus is an amphibious mammal about 13 feet long and up to 8800 pounds in weight. It lives in herds of 20 to 100 individuals dominated by a male. It spends the day sleeping or resting in the rivers of Africa. It leaves the water at night to graze on fresh grasses in the surrounding meadows. The hippo can hide almost entirely under the water, keeping just its nostrils, eyes, and ears above the surface. Thus, it can smell, see, and hear everything that happens on the surface without being detected. It is like a fish in water, but it can be very agile on the ground, too. The hippo can reach a running speed of 18 miles an hour. It has enormous ivory teeth in its lower jaw that it shows regularly when yawning. These teeth are really fearsome weapons used by the males in their fights during the mating period. Hippos mate in water. The female gives birth to one calf outside of the water, but soon enough the mother and her calf start for the water. The calf suckles under the water.

THE GRIFFON VULTURE

The vulture is one of the largest birds of prey. In fact, there are 15 separate species of vultures. The griffon vulture is the largest and the strongest of all its kind and can be seen in the picture above. It has a wingspan of almost 10 feet and weighs about 17 pounds. The griffon vulture is a champion glider. It is capable of gliding for hours at an altitude of 11,500 feet, without flapping its wings. It uses warm air currents to stay in the air. It is a carrion feeder and, because of its sharp eyesight, it can rapidly discover the smallest carcass from an impressive distance. It is also the first at the scene of a tragedy. Its long, usually bald neck allows it to rummage through a carcass, shredding the flesh with its strong hooked beak. It travels long distances and does not settle in a certain place except for the mating period. The female lays just one egg. The eaglet will be able to fly starting at the age of four months.

The leopard is a very good acrobat. This robust, solitary cat spends a great part of the time in trees, surveying its territory of more than 40 square miles. When night falls, it leaves its shelter and goes hunting. Gazelles and baboons are its favorite prey. It has few enemies; just lions, hyenas, and large baboons dare to attack it ... and man too, who hunts it for its fur. Did you know that the leopard and the panther are one and the same animal? Some leopards are black, like Bagheera from *The Jungle Book*. Others are almost white, like the snow leopard. When the female is ready to mate, she produces special urine that attracts males. Fights begin among the males, and only one winner emerges in the end. Four months later, the female will bear one to six cubs. They leave their mother at about two years of age to start living alone.

THE LEOPARD

THE AFRICAN BUFFALO

The buffalo is the biggest bovine of Africa. (The male can weigh 1800 pounds.) It lives in herds sometimes composed of hundreds of animals. It is found especially in swampy regions. The buffalo loves to lounge and to roll in the mud. It is one of the most dangerous animals you can meet. Its curved horns are very impressive. Some males have horns that measure more than three feet across . . . and they know perfectly how to use them. It is said that in Africa more hunters are killed by buffaloes than by any other wild animal. Man has never succeeded in domesticating it, not even by crossing it with cattle. The African buffalo has no natural enemies, other than man and the lion. The hyenas and crocodiles swarming in the swampy areas do succeed in catching a young buffalo from time to time. The female gives birth to one calf at a time. The buffalo lives an average of 25 years.

THE MARABOU

The marabou is a large wading bird. This funny-looking bird is five feet tall, and its wingspan may reach 10.5 feet. It belongs to the stork family and is also called the "stork with a pouch." This nickname comes from the pouch hanging from its neck, functioning like a meat safe, where it stores food left over from meals. You may find it on top of a tree or near a stream. The marabou is a very good fisher, and fish comprise its main diet. It also plays the role of a garbage collector; it doesn't hesitate to help eagles clean the leavings of prey after lions and hyenas have dined. You may also meet this bird in villages, where it feeds on food wastes and cleans the streets of garbage.

The warthog is the African relative of the wild boar and lives in small groups, called "sounders," of 10 members. It is usually gray or black, but since it likes to wallow in the mud, it often becomes red or yellow. When running, it lifts its tail up in the air, which makes it look rather funny. Its enormous head is armed with two pairs of tusks. They are, in fact, highly developed canines; older warthogs may have 24-inch-long tusks. The warthog uses them to dig out the plants, tubers, insects, and larvae it feeds on. It must kneel to dig out its diet, but it must be very careful while in this position, because it is then highly vulnerable and exposed to predators (hyenas, lions, or panthers). As you can see in the picture opposite, the male differs from the female by having

two pairs of big warts on his "cheeks." After a five-month gestation period, the female uses an aardvark burrow (the old den of an anteater), where she can bear three to five young.

THE WARTHOG

THE DIK-DIK

Being 24 inches long and 14 inches tall, the dik-dik is one of the smallest antelopes in the world. Its name comes from the distinctive call that it produces when worried. It can also produce very sharp sounds. It is a shy animal that starts running in zigzag leaps at the slightest suspect noise. Females are hornless. Because dik-dik meat is not as desirable as the meat of other antelopes, hunters don't disturb them. When chased by predators, the dik-dik can easily sneak through a thorny thicket because of its small size, which often discourages its chasers. This is why this little antelope lives in the prickly regions of the savanna. Despite this advantage, the dik-dik is very prudent and waits for nightfall before going out for food. It feeds on grass, acacia leaves, and fruit, which provide an important part of its water supply.

THE HYENA

The hyena is probably the least pleasant animal of the savanna. Its treacherous appearance, its reputation as a coward, and its carrion-eating habits contribute to its lack of popularity. Its yell sounds like horrid laughter, and from this it has gotten the nickname of "laughing hyena"; it can also bark, snarl, and even imitate the roar of a lion. Its diet consists mainly of the leavings of animals killed by lions and cheetahs, for which it must fight with the eagles. Its jaws are incredibly powerful; therefore, the hyena can easily break the bones of a zebra or a buffalo to get at their marrow.

Sometimes it hunts by itself; it hunts in packs when night falls, chasing its prey for very long distances. Sometimes, it can even attack lions! Unlike other animal species, female hyenas are larger than males, and the clan is always headed by an old female.

THOMSON'S GAZELLE

This very delicate little gazelle lives in large herds, just like many other antelopes. It is permanently on the alert; it can see danger from more than 330 yards away. When it senses the slightest threat, it starts running away in long, zigzag leaps. It is probably the fastest antelope and can run about 50 miles per hour. However, it often falls prey to lions, cheetahs and hyenas. When the grass is green, it has all the water it needs. But when there is a drought (as often happens in Kenya and Tanzania), it uses a number of adaptations to conserve the water in its body: its light-colored fur reflects the sun's rays, it doesn't sweat, its excrement is very dry, and its urine is highly concentrated. Moreover, it feeds at night (when it's less hot and the plants are covered with dew); in the daytime it ruminates while standing still. Each of its horns consists of a bone attached to the skull and covered entirely by a horn-like integument; the horns grow throughout the gazelle's life, and they never are shed.

THE RHINOCEROS

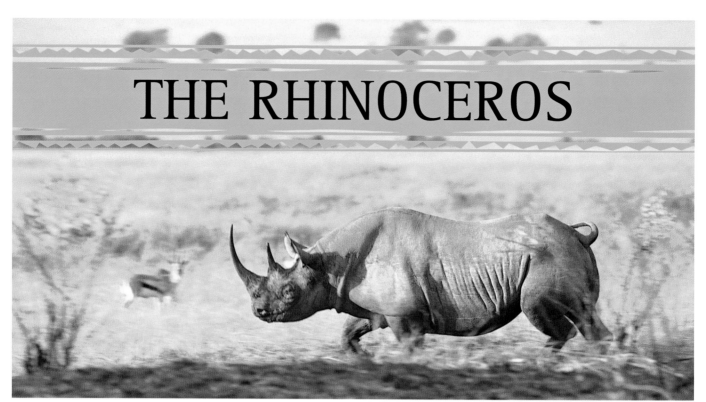

In terms of size, the rhinoceros occupies second place in the category of terrestrial animals (immediately after the elephant, which is in first place, of course). However, despite its 6,600 pounds, it is very agile and can run as fast as a horse. It lives in a family called a "cash" of three or four members, which travel slowly across the savanna and eat grass. It has poor eyesight, but it has excellent smell and hearing. The Asian rhinoceros has one horn on its nose, while the African rhinoceros has two. Each of these two horns may be 60 inches long. They differ from the horns of a cow in that they are made of the same matter as our nails. To procure these horns, poachers have almost exterminated the rhinoceros.

These horns are really very expensive; they are used in the production of all sorts of objects and medicines demanded in Asia and Africa. The rhinoceros's skin is very thick and protects it like a shield.

The rhinoceroses from India have "slate-like" skin, creating the impression that they are covered in real armor. The rhinoceros is not afraid of other animals, just of man.

THE SABLE ANTELOPE

The sable antelope is one of the most beautiful antelopes, but it is also one of the most aggressive. It has two long, strong, ringed horns curved backward. It is also called the "horse antelope" because of its brush-like mane and because it's the size of a horse. The largest may reach 600 pounds. Like all antelopes (there are about 150 species), the sable antelope is a cud-chewing mammal whose horns never are shed. Both the male and the female have horns and use them mainly for defense. The sable antelope is a fast animal, living in herds of eight to 40 led by a male. The herd is nomadic; it moves constantly in search of food, namely plants. The lion is its most important enemy; many people believe that the lion is the only animal capable of killing it of assuming the risks involved in trying to kill it, of course! The sable antelope is a courageous animal that defends its life stubbornly even against a lion.

THE JACKAL

There are several species of jackals. The common jackal presented in the picture lives in the north of Africa, in the southeast of Europe, and in the south of Asia. The jackal is related to the coyote, which lives on North American plains and deserts. Also called the "wild dog," the jackal is restless; it can run for hours, covering long distances. It hunts alone or in a pack, howling all the time. It feeds on rodents, eggs, insects, and small antelopes, and whenever necessary, it can feed on carrion too. The jackal is as cunning as a fox. Whenever it notices a young prey, it tries to isolate it from its mother to attack it easily. The jackal has a highly developed sense of smell; when it feels that a predator is near, it produces a warning yell that all the animals in the surrounding area benefit from. In spring, the female gives birth to as few as two or as many as 12 pups sheltered in a den.

THE LECHWE

Lechwe are large-sized antelopes that weigh up to 265 pounds and can be more than 60 inches long. It is quite a rare species, yet it can be found in almost all regions of Africa. It lives in large herds, usually near a water source (a river or a swamp) that it can easily cross by walking or swimming. Only the male has a pair of characteristically ringed horns. He uses them to defend against predators. During the mating period, he uses his horns while fighting for a position in the hierarchy of other males in the herd. These are usually ritual fights, with little violence. The strongest male chooses the females that will form his herd. Each female gives birth to a single calf after a seven- or eight-month gestation period. The mother weans the cub by the time it's four months old. The females become sexually mature at two years of age and the males at three.

THE IMPALA

he impala is the most rapid and elegant antelope. It is distinguished by the black stripes on its hind legs, as you can see in the picture below. The females are hornless. The impala can gallop up to 37 miles per hour. It can leap 10 feet high and a distance of 11 yards. In case of danger, these wild leaps can confuse predators (lions, leopards, and cheetahs), and sometimes allow it to escape. For almost the entire year, the males group alone in small herds. They rejoin the females to mate at the beginning of the rainy season when the grass abounds. During this period, the herds consist of a male accompanied by between five and 50 females. The young calves stay with their mothers for one year. They are often found in the woodlands of the savanna, always near a water source.

THE BABOON

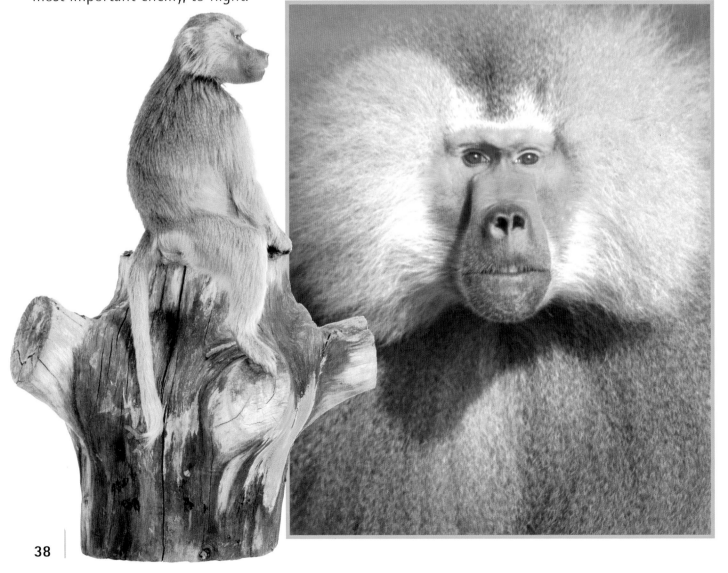

The baboon lives in a hierarchical colony, composed of from 30 to 100 individuals. It is an intelligent animal with a great capacity for learning. If the baboon sees a tent, it doesn't hesitate to enter and steal something before disarranging everything around. It particularly likes the rocky regions of the savanna. Baboons feed on plants, fruit, insects, small reptiles, and sometimes they even hunt small antelopes. Their bodies are adapted to terrestrial life; they walk leaning on their soles and palms. (This is why we call them "plantigrade" animals.) Males are twice the size of females; they may stand three feet tall and more and can weigh 88 pounds. In baboons, the gestation period lasts six months; after that, the female gives birth to an infant that clings to its mother's belly almost indefinitely. The baboon is amazingly powerful and can be very aggressive; thus, the baboon can put the panther, its most important enemy, to flight.

THE GNU

Although not as graceful as impalas, gnus belong to the antelope family, too. Its general appearance is rather funny. Imagine the body of a gazelle and the huge head of a buffalo with strong, broad horns curved upward. The gnu antelope lives in herds of thousands of members. Numerous populations are still found in East Africa, especially in Kenya and Tanzania. The herds migrate depending on the season to find the best pastures. This is what we call "transhumance." Their trips are long, and nothing can disturb their course, not even the large rivers swarming with crocodiles. Like all the antelopes of the savanna, the gnu is a fearful, peaceful animal. However, to protect its young, it doesn't hesitate to confront a cheetah, fighting and kicking viciously with its hooves. Its most important predators are lions, hyenas, and wild dogs.

THE AFRICAN WILD DOG

At almost three feet long, this animal resembles a dog and a fox at the same time; just like these two animals, it belongs to the dog family, known scientifically as "Canidae." It lives in groups of seven to 10 adults, led by a dominant pair. Its hunting technique is very efficient: it hunts in packs, each member chasing the prey (mid- or small-sized antelopes) in turn until the prey is exhausted. In the end, the prey is surrounded and killed. After more than a two-month gestation period, the dominant female bears six to 10 pups in a den. The pups are brown at birth, but after a while, their fur becomes flecked with yellow, black, and white just like their parents' fur (each pup can be identified according to its spots.) The African wild dog is an endangered species; it is sensitive to the diseases of domestic dogs and is hunted by local inhabitants who try to protect their flocks of sheep.

THE PORCUPINE

The porcupine has enormous spines, or quills; the longest quills are on its back and can be 10 inches long and 0.2 inches in diameter. Despite its "thorny" look, this rodent is completely harmless. It goes out at night to find plants and fruit to eat. It lives in North America, South Europe, Asia, and Africa. When it feels threatened, it turns around and moves backward. Its spines detach easily; thus, if they touch an opponent, they remain stuck in its skin, provoking severe pain. The only way its predators (such as the lynx) can eat it is by turning it over on its back and catching it by its spineless stomach. The porcupine moves with difficulty on the ground, but it is an excellent swimmer. It has strong claws with which it digs a den to sleep in during the daytime. After a two- to four-month gestation period, the female gives birth to a baby. Baby porcupines are called porcupettes, and they can live up to 20 years.

Under the Desert Sun

THE DROMEDARY

romedaries are cud-chewing animals that live in North Africa and the Middle East; there are also several thousands in Australia (brought there by man before the end of the 19th century). Dromedaries closely resemble camels, their relatives from Central Asia. However, you can distinguish them very easily: dromedaries have one hump, while camels have two. These humps are actually fat reserves that allow dromedaries and camels to survive for several days without food or water. The dromedary is very well acclimated to desert life: its legs are adapted so that it won't sink in the sand, and it is able to endure the heat. The dromedary feeds with no difficulty on thorny plants growing in the desert. It can also close its large nostrils to keep the sand out during sand storms. The dromedary is a domestic animal; it is used as a riding animal (to carry people) or as a beast of burden (to carry loads that may weigh 1,000 pounds).

THE RATTLESNAKE

The rattlesnake is the American relative of the viper; it can reach a length of 60 inches. As is true of all reptiles, the rattlesnake sloughs its skin, that is, it regularly changes its skin, which is made up of thousands of scales. The "rattle" at the end of its tail consists of thick skin rings. At birth, the young rattlesnake has only two embedded rings. The others appear after each sloughing and form a kind of noisemaker rather soon. When the rattlesnake is troubled, it lifts and agitates its tail, making the rings vibrate. This jingle produced by its tail - from which the "rattlesnake" gets its name - is heard from far away and warns an intruder that he best stay away from its poisonous fangs. When this sound is heard, the rattlesnake can be 20 or 30 yards away. The rattlesnake is mostly a nocturnal animal; it stays in the shadows in the daytime and goes out hunting when night falls. It doesn't hear, but it can locate its prey (rodents, birds, frogs) because of the heat their bodies produce. Its venom is very dangerous and kills the prey rapidly. (It can be fatal to humans as well.)

THE FENNEC

his large-eared little fox has a very nice look, and it can also be tamed. But don't forget that the fennec is a wild animal, a real carnivore, very agile, and can be very aggressive, too. It hunts at night; it feeds on small rodents, birds, lizards, and insects (especially locusts) and occasionally it adds fruit to its menu. It spends the day in a den, usually dug in the sand, sheltered from the torrid heat of the desert. This den consists of several galleries and a room lined with plants, fur, and feathers, used as litter. Here it lives with its mate or family. Following about 50 days of gestation, the female gives birth to two to five cubs once a year.

THE MONGOOSE

The mongoose is 3.3 feet long, including its tail. It is very agile and can become a redoubtable opponent. It is persistent and never releases its prey. The mongoose is also the most famous enemy of the fearsome cobra. It avoids the fangs of this snake by stepping aside rapidly from its attacks until it manages to leap and catch the cobra just behind its head. Its skin and thick fur, highly resistant to venom, protect it. During the age of the pharaohs, mongooses were held sacred by the

Egyptians; mummified mongooses have been found in the pyramids. The mongoose likes to eat crocodile eggs so much that it prevented these dangerous animals from invading the land bordering the Nile.

THE ORYX

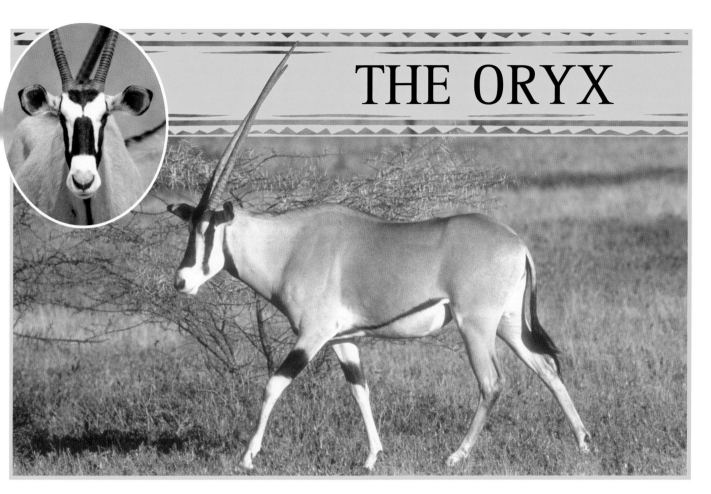

This superb antelope, with black, spotted head and legs, is highly resistant to fatigue. It can walk for 18 hours in search of food without a break, covering more than 60 miles. The oryx feeds on grass, fruit, and roots, which provide a good quantity of the water it needs. Its most important enemy is the panther. But pay attention: attacking an oryx is a very dangerous action! It is indeed capable of defeating its enemies, using its long, thin, sword-like horns (which may be more than 3.3 feet long). Oryxes live in herds of 10 to 60, where the females and the males are equal in number. The herd is very important. If a male is moved away from the herd, he will come back to the place where the herd was last, waiting alone for its return.

THE SCORPION

espite its 12 eyes, the scorpion has bad eyesight. Fortunately, it doesn't need good eyesight too much, because it is a nocturnal animal. It hunts at night and is capable of detecting its prey (spiders, insects), because it has fine hairs that are highly sensitive to vibrations. When catching prey with its pincers, the scorpion immobilizes its victim by injecting venom into it with the stinger at the end of its tail. The stings of scorpions resemble those of bees. Therefore, contrary to common opinion, scorpions are not generally dangerous to humans. But note that the sting of certain scorpions, such as those living in the Sahara, may be fatal. During the mating period, the male and the female use their pincers to stay close together; the male turns the female over to make her pass over the semen he has spread on the ground. After birth, the young scorpions shelter themselves on their mother's back for 10 days. Depending on the species, the length of a scorpion may vary from one to seven inches.

THE MOLOCH

This little reptile (a lizard of about six inches) looks like a real monster coming from an old legend meant to disturb children's sleep. It truly deserves the name of spiky moloch. Despite its appearance, it is tame and harmless. You can even safely hold it in your hand. Its wide, flattened body is covered entirely with spines. It has two spines on both sides of its head that are much stronger than the others. These spines look like a pair of horns. Its predators must have a hard palate to eat so

many spines. The moloch is well adapted to desert life: in the morning dew condenses on its spines and flows into its mouth. Thus, it can live for weeks without finding a water source. It buries itself in the sand to escape the desert heat. It feeds especially on ants, and just like chameleons, the moloch can change the color of its skin according to its mood.

THE SPRINGBOK

The springbok is very common in the south of Africa. This gazelle lives in herds of 10 to 200, and sometimes even more. It holds the record for the long jump: it can jump 16 yards. It can live without ever drinking water: its varied diet (fruit, flowers, plants, bulbs) provides all the water it needs. It is also the only gazelle that can eat several toxic plants. When a predator is near, the springbok is often the one to give the alarm by jumping up in place, legs stiffened and back bent. (These jumps can be 10 feet high.) The springbok can't rely on its small horns to defend itself, and therefore it prefers to run. Besides, it can disappear incredibly fast (40 miles per hour) to get away from hyenas and lions. Every two years the female gives birth to one cub (rarely two) after a four- to six-month gestation period. The mothers form a "nursery" within the herd and tend the calves together.

THE COBRA

Cobras are the longest poisonous snakes; the royal cobra holds the record, at 20 feet long. When disturbed (like the snake in the picture below), the cobra rises from the ground and spreads its "hood" (loose skin on the neck) in a very impressive manner. The cobra's venom is extremely strong, so that its bite can kill even an elephant. Some cobras, like the one in the picture on the left, can spray venom through their fangs from a distance of about 80 inches, aiming always at the enemy's eyes to cause temporary blindness. Cobras feed on lizards and other snakes. They have few enemies; in fact, only mongooses and secretary birds manage to kill them. Cobras are the Indian snake charmers' favorite. The charmers succeed in making the cobra rise up slowly and sway by playing on a pipe. Actually, the snake can't hear the music (all snakes are deaf); it only follows the movements of the charmer's pipe.

THE STRIPED GROUND SQUIRREL

he striped ground squirrel is smaller than the common red squirrel and lives in India. Like all other terrestrial squirrels, its tail is not very thick, and its front paws are very strong, enabling it to dig in the ground. This squirrel lives in a den dug all by itself. Unlike tree-dwelling squirrels, terrestrial species hibernate, and therefore they spend the winter sleeping in their dens. Depending on the species, the gestation period lasts from 24 to 44 days, and the number of pups varies from three to eight at each birth. The striped ground squirrel is a vegetarian - notice it in the picture opposite, busy crunching a big seed. However, it occasionally eats a meal of insects, eggs, or even small birds. Sometimes it can do quite serious damage to cereal crops.

In the Rainforest

THE BOA

This snake can be almost 15 feet long. It has a bifid tongue (forked or divided in two at the end), which it uses as a nose to "taste" the surrounding smells. It is a solitary nocturnal animal. In the daytime, it shelters among rocks or up in trees, like the boa snake from Madagascar seen in this picture. The boa is a ovoviviparous snake. This means that the female keeps her eggs in her genital tract till they hatch, giving birth to live young, already 20 inches long, generally numbering from 20 to 50. The boa is not a very good hunter, but fortunately it can live several weeks without food. It feeds on large lizards (like iguanas), birds, and mammals (rats, mongooses, squirrels, ocelots). It is not a venomous snake; therefore it kills its prey by winding its body around it and suffocating it. Usually the boa doesn't attack humans, unless it is hungry or it cannot escape.

THE CHAMELEON

The chameleon is a reptile whose length varies from three to 24 inches, depending on the species. It is considered a master of the art of camouflage; it can change its color depending on its environment (to go unnoticed by its prey or predators) and its mood (for example, to show that it is angry). The female lays eggs that hatch after six-to-nine months. The eyes of the chameleon are particularly distinctive: they can move independently of each other in many directions, thereby permitting a very wide field of vision. Imagine that it is capable of fixing on its prey with one eye and surveying its surroundings with the other to make sure there is no predator around... and all this without even moving its head! Insects are its favorite prey. Once an insect comes into sight, the chameleon fixes it with both eyes to measure the distance; then it darts its long, sticky tongue (when unfolded, its tongue is longer that all the rest of its body) into the air with rapid and extraordinary precision, catches the insect, and swallows it whole.

The chimpanzee lives in family units and is a gregarious animal that establishes highly developed social relations. To say hi, it touches whomever it's greeting with its fingertips and offers a kiss or a friendly pat on the back. It is first in the classification of animal intelligence: it uses tools (for example, it uses a rock to break nuts), throws stones at its enemies, and drinks water from a glass made of leaves. It lives primarily in trees, but it can either walk or run on the ground on its hind legs, just like humans can. At night, the entire tribe gathers in a big tree, and all the members prepare their shelter by using branches. Starting from the age of five or six, females can bear infants, and the gestation period lasts nine months, the same as in humans. Although the young infant is cute and funny, adores playing, and does a lot of tricks, remember that the adults are very powerful animals, and you must be careful with them. (A male can be 5.5 feet tall when standing and can weigh up to 154 pounds!) You can find them in zoos, usually on a small island. And do you know why that is? Because they can't swim, and therefore they can't escape.

THE CHIMPANZEE

THE GIBBON

This ape can stand about 3.3 feet tall. The gibbon has very long arms, making it very agile in trees; it easily passes from one branch to another by swinging. In contrast, when it is on the ground (which happens rather rarely), it usually keeps its arms raised while walking, using them to keep its balance.

It feeds on plants (fruits, leaves, and flowers), insects, and small birds and their eggs. Just like the chimp, the gibbon has no tail. It lives in small families and defends its territory by howling (like the one in this picture is doing) or by fighting with intruders. The gibbon is a monogamous animal (which means that it has a single mate); the couple has only one infant, which will stay with its family till the age of five or six. The gibbon is the only ape that doesn't build a shelter to sleep in. It spends the night on a branch, crouching near its mate.

THE GORILLA

The gorilla is the largest of the great apes. It weighs up to 660 pounds and can be 7.5 feet tall. When disturbed, it stands up and howls, beating its chest with its fists to impress its opponent. However, despite its extraordinary physical strength and its general appearance as a ferocious beast, the gorilla is rather shy. We now know that it withdraws when meeting humans, unless it is threatened. Unlike other animals, the gorilla rarely climbs trees. It walks on the ground mostly, using its four paws. It lives in groups consisting of a large male, three or four females, and their infants. The gorilla spends most of the time looking for leaves, flowers, roots, and bamboo shoots to eat or lazing in the sun. Every night, when preparing for sleep, the gorilla builds a nest directly on the ground from branches. Humans are its main enemy. Trophy collectors have hunted gorillas for quite some time. Presently, gorillas are protected, yet poachers still claim numerous victims.

THE IGUANA

The iguana is a giant lizard, which may reach a length of almost five feet. It has a crest of thorns all along its body, from its head down to the end of its tail, which makes it look like a dragon. The iguana is a vegetarian and therefore has numerous small teeth with which it can eat plants such as cacti. Occasionally it feeds on worms and insects. Despite its large size, the iguana is a tree-dwelling animal (which means that it can live in trees). You can often find it near a water source, jumping in the water at the slightest sign of danger. The iguana regularly leaves the trees to protect its territory against other iguanas. The females are usually more aggressive than the males, especially when it comes to choosing places to make their terrestrial nests. Despite all of this, the iguana is a rather gentle, harmless animal that you can easily tame.

THE KINKAJOU

This tiny animal is often called the "little bear." In reality, it is not a bear at all, but a relative of the rat. It differs from the bear by its small size, comparable to the size of a cat, and by the length of its tail. Its tail is prehensile, which means it's used to cling to branches. Moreover, the kinkajou spends most of its time in trees. It has a long tongue (about five inches), with which it can lick the nectar from flowers with long corollas. The kinkajou plays an important role in the plant reproduction process: while it is eating the nectar of flowers, pollen sticks to its fur, and thus the kinkajou transports it from

one flower to another. This process is called "pollination." The kinkajou also feeds on fruit, insects, bird eggs, and frogs. It is also very fond of honey and is considered one of the main enemies of wild bees. The enemies of the kinkajou are the margay, the jaguar, and the fox.

THE KOALA

The koala looks like a big, funny-looking, gray stuffed animal. It is a peaceful, harmless animal that sleeps 18 hours a day on the average. In the Australian natives' language, the word "koala" means "the animal that doesn't drink." The koala is an excellent climber and can stay in trees without coming down for years. It eats so many eucalyptus leaves (sometimes up to two pounds a day), that it smells just like eucalyptus. If you closed your eyes and sniffed a koala, you could almost mistake it for a sore-throat pill. The koala is a marsupial, just like the kangaroo. The koala cub stays in its mother's abdominal pouch for six months. After that, the mother carries it on her back for one more year. At the beginning of the 20th century, the koala was hunted for its silver-gray fur. Nowadays, it is a protected, but still endangered, species.

THE RING-TAILED LEMUR

These little, funny looking animals belong, scientifically, to the Lemuridae family; they are a kind of mammal that closely resembles the monkey. They are confined to the island of Madagascar. The ring-tailed lemur spends most of its time in trees and moves by jumping from one branch to another. When walking on the ground, it moves like a feline: its tail raised and moving from left to right. There are several species of ring-tailed lemurs, the most famous of which is scientifically classified as Lemur catta, the one depicted in these pictures. As you can see, the ring-tailed lemur has a very beautiful, long, black-and-white ringed tail that it uses to maintain balance when jumping. It lives in family groups. It is herbivorous, but it can also feed on insects or lizards. Although this species is protected by law, the Malagasies, who appreciate the ring-tailed lemur for its meat, constantly hunt it; the ring-tailed lemur is therefore an endangered species.

THE ORANGUTAN

The orangutan lives in the swampy jungles of the Indonesian islands of Borneo and Sumatra. In the Malay language, the word "orangutan" means "man of the jungle." Its thin, brownish fur represents the typical feature of the orangutan. It is very agile when swinging from one branch to another, but it is not very graceful on the ground. In fact, it can barely walk, because its bowed legs are short and have no heels. It is vegetarian, and its favorite fruit is the durian (a big, green, foul-smelling fruit). When night falls, it builds a platform from branches and leaves to sleep on. The female gives birth to only one infant, which will stay with her for five years. The young orangutans live in small groups after that, till they mature, between the ages of seven and 10. The adult orangutan is rather solitary, but sometimes it stays with others of its kind for several days.

THE TWO-TOED SLOTH

he two-toed sloth deserves its name. It spends at least 18 hours a day sleeping, and when it wakes up, it lives at a slow pace. It moves very slowly in trees, hanging suspended from branches, with its legs and face turned upward and its back downward. Its limbs are long and terminate in two or three long, curved claws, depending on the species. The sloth is so well adapted to its upside-down life that its fur grows in the opposite direction to that of other animals (from the belly to the back). Its fur has a strange greenish tint, which helps very efficiently to hide the sloth. This color comes from microscopic algae that grow in its fur. The sloth is a phytophagous animal (which means a plant-eating animal: shoots, fruit and flowers). Since its eyesight and hearing are not very good, it uses the senses of smell and touch to find food. The sloth mates in trees. The female gives birth to one baby after a six-month gestation period.

THE PARROT

The parrots you can see in these pictures are macaw parrots. The macaw is the largest parrot of all: including its tail, it averages 40 inches in length. Macaws are a rare species; therefore, capturing or taking one out of its natural environment is prohibited. The macaw lives in trees and is one of the most important characters of the rainforest! It is incredibly noisy, producing all sorts of cries. It is extremely curious and inspects everything around. It is also very skillful at imitating sounds: it is capable of repeating words or of reproducing the sound of a phone. It feeds on fruit, insects, and seeds. It has a very strong beak with which it breaks nuts and hangs on branches. When the male sees a female, he courts her by displaying his multicolored plumage. The female understands that he wants her to make a family with him ... and not with anyone else, because the macaw is very faithful, and the pair is formed for the rest of their lives. Macaws live in Panama and in South America and they are considered an endangered species.

THE PYTHON

The python is a close relative of the boa. Just like the latter, the python's fangs are not venomous. The python is most active at dawn and at dusk. It kills its prey by wrapping around it and coiling its rings to suffocate the victim. Its most common prey consists of small rodents, like rats and jumping mice, but it is able to kill larger animals as well. It is said that a python can kill and swallow even a leopard! Such a lunch would satisfy it for several weeks. The largest pythons can reach lengths of over 20 feet. When threatened, the python defends itself either by slithering away, by fighting, or simply by staying tightly coiled. The python is oviparous; this means that the female lays eggs. Young snakes shed their skin between four and six times a year; adults shed their skin no more than one to three times a year. The python sheds its skin from head to tail in one single piece, in the same way you would pull off an enormous sock. This sloughing process may last about 10 days.

THE ANTEATER

The anteater, of which the giant anteater is the most famous, is a strange animal. It is related to the sloth; both of them belong to the family of insectivorous mammals. The anteater is a genuine vacuum cleaner of ants and termites. The anteater rips open nests with its strong claws and then introduces its sticky, 16-inch-long tongue. It can swallow up to 30,000 ants a day. What is very striking about this animal is its long, spindle-like head, ending in a very small mouth with a sliding sticky tongue. The anteater is peaceful, but if threatened, it defends itself vigorously by using its claws. It sometimes can suffocate its enemies (pumas or cheetahs) by introducing its long tongue into their nostrils. After a six-month gestation period, the female gives birth to one baby, which will stay with her for the following two years before it starts living on its own. All this time, the mother carries her baby on her back.

THE TAPIR

The tapir is such a strange animal! It has a big belly like a pig, a small crest on its head, and a small but very mobile trunk. It uses this trunk to grasp the branches or the plants it feeds on or to breathe while swimming. It is a peaceful, shy animal, living generally in small troops. It is active at night, sleeping in the daytime and looking for food during the night. Its babies have broad, beige stripes and big, white spots all over their bodies and resemble older wild boar piglets. When hunted by a predator (pumas, tigers, or panthers), the tapir tries to escape by hiding in dense vegetation. Humans value the Brazilian tapir that you see in these pictures for its meat. Some populations in South America use certain organs as medicine; they believe, for example, that the heart of a tapir can heal epilepsy.

THE TOUCAN

The toucan is an imposing bird that lives in Central and South America. It is very colorful and has an enormous, distinctive beak, which contains air pockets and therefore is not as heavy as it seems. The function of this beak is not known for sure, but some specialists believe that it is a mere instrument of intimidation among the males. The toucan lives in groups of about a dozen. It is a very noisy bird, and it fully contributes to the boisterous atmosphere reigning in the rainforest. It feeds on pulpy fruit, seeds, insects, and spiders. During the mating season, the male and the female pass or throw berries from beak to beak. The male and the female mate and stay together for several years. The female lays two to four eggs in a nest once a year. Both parents incubate the eggs, and the chicks appear after 15 days. They will leave the nest after eight weeks. Natives love the toucan very much. They tame this bird just like the parrot and manufacture accessories and ornaments from its feathers.

THE TIGER

This superb animal is the largest of the cat family: it can weigh almost 660 pounds and can reach a length of more than 10 feet. Therefore, it fully deserves to be considered the most fearsome predator of the jungle. Generally, the tiger is not as fast as its prey, but it compensates for this shortcoming by relying on stealth when stalking its prey and by attacking without warning. The tiger is the only feline that sees colors. The tiger is not afraid of water, as you can see in these pictures. The

male lives in a wide hunting territory that he defends fiercely. The female gives birth to two to four cubs after a four-month gestation period. The cubs stay close to their mother for several years in order to learn to hunt. The tiger has no enemy other than man, who almost exterminated it by the beginning of the 20th century for the mere pleasure of collecting trophies. Presently, tigers are protected by law, but there are fewer than 5,000 tigers in the world, and poaching still takes its toll. The fur of a tiger costs a fortune, and various parts of its body are used in traditional Chinese medicine.

THE TORTOISE

The tortoise is one of the most ancient animals in the world; it existed long before dinosaurs! Its carapace of bony slabs is very solid and serves as a shelter. Whenever threatened, the tortoise can almost entirely withdraw into its carapace. However, the carapace is very heavy and doesn't allow it to move fast. The one you can see in these pictures is a giant tortoise. This species lives only in the Seychelles Archipelago, on the Aldabra Islands, and in the Galapagos Islands. Terrestrial tortoises can easily be distinguished from their aquatic relatives: their carapace is rounded and their limbs are thick, nonwebbed, and armed with claws. The tortoise doesn't have teeth, but it has a sharp beak with which it tears the plants and fruit it feeds on. During the reproduction period, the female digs a hole to lay her eggs and then she leaves. The adult tortoise can reach a length of 47 inches and can weigh more than 480 pounds. It is also the champion of longevity: sometimes it lives more than 150 years. Could anyone ever want more?

THE MARGAY

Did you know that wild cats still exist? Among them, the margay is very widespread in Central and South America. Its gorgeous spotted fur is the reason for its nickname of the "cat-tiger." Its big eyes are very well adapted for nocturnal vision. It is a very good climber; the arboreal life (living and sleeping in trees) is one of its characteristics. If you look at a domestic cat descending from a tree, you will notice that it steps carefully, keeping its head up. The margay is capable of descending from trees at full speed, its head forward, just like squirrels do. Its prey consists of monkeys, opossums, and birds. After a three-month gestation period the female gives birth to one or two kits, which are weaned at about two months. If man destroys its natural habitat (the rainforest), the margay cannot adapt to other environments, and attempts to have it reproduce in captivity have all failed. For this reason, the margay is considered an endangered species.

THE TARANTULA

The appearance of a tarantula is very impressive: its body can reach four inches in length, and it has four pairs of hairy legs. Just like the majority of spiders, the tarantula produces silk with which it covers its hole, which it digs in the ground. This hole has a kind of "lid" with a hinge. When the tarantula is disturbed, it hangs on this lid from the inside of its hiding place and stops the enemy from entering. The tarantula eats a great deal; it can stuff itself with food until the skin of its belly bursts. It waits for prey (small mammals and reptiles, little birds and insects) in its den; as it senses one of them passing by, the tarantula rushes out of its hiding place, jumps on the prey, and bites it with its two fangs, injecting venom that paralyzes the little animals. During the mating period, the male constantly knocks on the lid of the female's hideout. When one comes out, the two come close and copulate. The female lays about 200 eggs that she protects in a silk cocoon.

THE BAT

The bat is the only mammal capable of flying. Its wings are made of thin skin stretched between its body and its long fingers. In the daytime, hundreds or thousands of bats gather in caves or in trees to sleep, their heads downward; at night, they go out in search of food. The bat generally feeds on insects; some of them feed on fruit and flower nectar, too. Certain bats feed on the blood of other animals; these are called "vampire-bats." To orient itself in the dark, the bat uses a radar system, uttering short, but extremely sharp yells (that we can't even hear). If an obstacle or a prey reflects the sound, it produces an echo that is picked up by the big, very sensitive ears of the bat. This way it is able to avoid obstacles or to detect and follow its prey. The female bat gives birth to one pup at a time; she carries it on her back until it can fly alone.

THE PANGOLIN

ust like the sloth and the anteater, the pangolin belongs to the family of edentate (toothless) mammals. Its body is covered with scales, and you can almost take it for an anteater dressed entirely in a shirt of scales. If you see it climbing in trees, you could mistake it for a pinecone. Its horned scales are of the same material as hair and are arranged on its body just like the tiles on a roof. Like the anteater, the pangolin feeds on ants and termites, using its sharp nose and sticky tongue. The pangolin lives in pairs and is especially active at night. It spends the day sleeping in its burrow, covering the entrance with soil so that it won't be disturbed. When it is attacked, the pangolin defends itself by rolling into a ball, thus protecting its unscaled head and abdomen. The pangolin is an endangered species. Unfortunately, the pangolin cannot count on successors to help it avoid extinction, because the female gives birth to just one baby at a time. The baby has soft scales that harden in the second day of life. The baby hangs on its mother's back for transportation, just like infant monkeys.

THE PRAYING MANTIS

ake a good look at this straight, 2.5-inch-long insect with two legs close to its head; doesn't it remind you of a nun in prayer? The praying mantis can stay like this for hours, completely motionless, lying in wait for its prey (flies, locusts, butterflies), which it catches by unfolding its abducting legs at full speed. Certain tropical species are very beautiful in color, looking like the flowers they hide in, ready to pounce on their prey. The female is always larger than the male, and her ferocious appetite makes her devour him while, or just after, copulating. After the insemination, the female lays eggs in an egg case (called an "ootheca"), containing hundreds of eggs. She covers them with whitish foam that hardens upon contact with air and protects the eggs from predators (lizards) and cold weather.

THE GECKO

As soon as night falls, the gecko raises its small, dragonlike head and starts hunting the insects it feeds on. The gecko can move very fast, in any position, even with its head downward, and cling to all kinds of surfaces, even smooth ones like glass. It can do this because of the wide, distinctive disks on its toes: its digits are covered by millions of tiny hairs, called "silk hairs," which act like suction cups. The smallest gecko is 0.4 inches long and the largest can reach a length of 14 inches. When walking on the burning sand, the desert gecko executes a sort of dance to cool its feet: it lifts them in the air in turn, and sometimes it stays on its stomach in order to raise them all at the same time. The gecko has another special feature: when it is in danger, it yells and it can even bite. Like many other reptiles, the female lays eggs (generally two), and the young hatch after a two- or three-month incubation period.

THE BLACK PANTHER

The black panther is one of the most mysterious inhabitants of the jungle. It is seen only rarely in its natural environment. Despite this, everybody knows this splendid animal, which has become famous because of characters like Bagheera from *The Jungle Book*. But don't let yourself be fooled. The panther is not a cute character in cartoons; it's a wild feline in every sense of the word. The black panther is mostly a nocturnal animal: it leaves its shelter at dusk to go hunting. It has no enemy other than man, who hunts it for its wonderful black fur. Although it looks like it's completely black, its fur is actually spotted, just like the fur of all panthers. But the spots are very subtle and can barely be seen.

In Mountains

THE ROYAL EAGLE

The royal eagle is a splendid, magnificent bird of prey, with a wingspan of more than 80 inches. It lives in pairs and nests in the walls of cliffs, or rarely, in the top of a tree. Its nest is made of branches and is covered with grass; sometimes it can reach 6.5 feet in diameter. The female lays and hatches generally one egg, sometimes as many as four. While the mother protects the chicks, the father hunts to feed them. To find prey (rabbits, marmots, lizards, sometimes even young black goats, foxes, or lambs), he flies over a territory averaging some 350 square yards. As soon as he spies his prey, thanks to his very penetrating eyesight, the eagle swoops down on the prey with breathtaking speed (more than 90 miles per hour!), grabs it in his claws, and takes it away. Royal eagles have long been accused of kidnapping children. This is impossible, of course, and yet, the royal eagle was once persecuted in Europe. Now this species is protected.

THE EUROPEAN MOUFLON SHEEP

The European mouflon sheep is a wild animal. It is larger and more intelligent than its domestic relative. It lives in flocks of 10, which easily wander the mountains and are capable of making great jumps. The sheep shown in these pictures live in Asia, but there are mouflons in Europe, Africa, and North America as well. The male and the female have spiraling curved horns; the male's horns are very long; they can measure more than 40 inches. Their sharp vision enables the mouflon to see predators (the wolf and the puma) from far away. In summer, the males (also called rams) fight one another with their long horns to join the flocks of females. Four months later, the inseminated females leave the flock to give birth to one or two lambs. The lamb starts grazing alone at two weeks and is weaned at six months. The mouflon can live 10 years, unless a hunter bags its head for his trophy room.

THE IBEX

The ibex is a large mountain goat, with two ringed horns that are curved backward (some males have hornes almost 40 inches long). It climbs and jumps from rock to rock with agility and awesome energy. You've hardly even noticed it, and it has already dashed away! It lives in flocks of 10 (females or males) and feeds on plants. In winter, the flocks descend from the tops of mountains and shelter in the woods. The ibex feeds mainly on lichens and mosses at this time of the year; sometimes it can dig out several wisps of grass by routing in the snow. The flocks of males and females meet and start mating in winter, too. The males fight each other with their horns to attract the females. By the end of spring, the female gives birth to one or two kids that climb with her toward the peaks of the mountains to spend the summer there.

THE ROCKY MOUNTAIN GOAT

This stumpy goat has very thick fur and short legs and lives in western Canada. It climbs the rocks skillfully. To see it, you must climb up to an altitude of at least 6,500 feet. Both male (the he-goat) and female (the she-goat) have small, sharp horns like daggers. The Rocky Mountain goat grazes in herds. The herd is headed by a female accompanied by her kids, except during mating periods, when a male takes over leadership. The skull of the he-goat is rather thin and fragile; during fights, he avoids attacking straight on with his head, as his cousins, the ibex and the mouflon, generally do. He quite often avoids fighting altogether and prefers to solve domination problems by parading. The main enemies of the Rocky Mountain goat are pumas, royal eagles (which attack the kids), bears (rather rarely), and man (who hunts it as a trophy kill).

THE CHAMOIS

The chamois has a broad, dark stripe on its back and another one at eye level. Although it is less agile on rocks than the ibex, to which it is related, this cud-chewing goat manages well enough. It lives in flocks of five to 30, generally at altitudes between 3,300 and 9,800 feet. It is capable of climbing from one level on a slope to another one 3,300 feet above in 15 minutes. It spends the summer in snow line areas. In winter, it descends to the woods to escape the wind and to satisfy its diet, which consists of plants, buds, mosses, and pine shoots. The rut period occurs between November and December. To mark his territory, the male spreads a very strong scent, which is produced by the glands situated at the bottom of his horns. In May, the female gives birth to one kid (very rarely two) after six months of gestation. Certain species of chamois are threatened by extinction. Indeed, these animals have been excessively hunted for the mere pleasure of hunting or for their soft supple skin from which the famous "chamois leather" is made.

THE LLAMA

This cud-chewing animal from the Cordelier Andes is related to the camel. There are, in fact, four species of llamas – two domesticated species and two wild species. The two domesticated species (the common llama and the alpaca llama) come from one of the two wild species, which is called "guanaco." The second wild species is called "vicuna." Llamas make ideal pack animals, because they can carry a burden of 110 pounds for 12 hours. These pictures depict the common llama (distinguished by its longhaired fur). The alpaca llama has soft, silky, shorter hair and is bred especially for its highly valued wool. Llamas usually live in harems. The female gives birth to one infant, called a "cria," after a one-year gestation period. When the llama spits, it does not do so from a lack of respect, as has been believed, but as a way of making its enemies stay away. Therefore, you best avoid provoking a llama, because it's not pleasing at all to receive a shower of llama saliva in the face. Icky!

THE PUMA

This gorgeous cat is also known as a "cougar" or "mountain lion." It can reach a length of 79 inches (including its tail) and can weigh almost 220 pounds. The puma can be found across the entire American continent, from north to south. It lives in almost all habitats: conifer and broad-leaved forests, mountains, swampy regions, rainforests, plains, and deserts. It holds the indisputable record for long jumping: it is capable of jumping 45 feet without even taking a step. Unlike other cats that roar (the lion, the tiger, the leopard), the puma purrs just like its relative, the lynx. After three months of gestation, the female takes refuge in a cave to give birth to from one to six cubs that she nurtures alone; she does not let other animals come close to the cubs, not even the male, for fear he might devour the cubs. The puma is very fast and agile; it can reach a running speed of 49 miles per hour. It hunts large animals, especially stags, but also mouflons, porcupines, or rabbits. Sometimes it attacks domesticated animals, which makes man fear and hunt it.

THE YAK

The yak is a cud-chewing animal that can stand as tall as 6.6 feet. It lives in the high mountains of Tibet and in northeastern China. You can meet a yak as high up as 19,600 feet, above the tree line. As you can see in the picture, its body is terribly large and entirely covered with long, silky wool that is white or black in color. Man has long domesticated the yak. The Tibetans use it for riding and as a pack animal and also as a source for wool, leather, meat, and milk (from which they make cheese); they even use its excrement as fuel or fertilizer. There are still a few wild yaks; they live in herds consisting mostly of females. The male lives alone most of the time and joins the herds of females only during the reproduction period (which is called "rut").

THE CONDOR

The condor is a bird of prey found across the entire American continent. The Andes condor has a wingspan of more than 10 feet. It is actually the largest flying bird in the world and can soar up to a height of 16,400 feet; its flight is very imposing. Not very attractive, the condor is a carrion-eating bird, like its relative, the vulture. Its sharp vision enables the condor to detect carrion from great distances. It lands just like a plane: it lowers its legs first, brakes with its wings at the last moment, lifts its tail a little, and eventually reaches the ground very slowly. You must climb up into the mountains to see it. Farmers hunt condors to protect their cattle, and therefore they are a very rare species. There are no more than 60 registered condors in California. Condors mate for life and breed very rarely. The female lays only one egg every two years. The young condor leaves its parents at two years of age; it reaches maturity when it is six or seven and can live for 65 years.

THE MARMOT

The marmot is a rodent that lives at an altitude of 5,000-10,000 feet. Starting in the fall, it hibernates for five months. It shelters in a burrow, and its body functions change: its cardiac rhythm slows down from 80 to four beats per minute, its respiration rate changes from 40 to two breaths per minute, and its body temperature goes down from 98° F to 41° F. In spring, the female gives birth to two to nine blind, furless young. They are weaned after five or six weeks, and they can live for five years. The main predators of the marmot are the wolf, fox, coyote, dog, brown bear, and lynx. Unlike the majority of routing animals, the marmot has very good eyesight. It is on guard all the time and keeps watch in a vertical position, near its burrow, where it takes refuge at the slightest sound. However, it doesn't give up very easily when caught and is able to face down a fox or a dog.

THE GIANT PANDA

You surely have seen a photo of this beautiful black-and-white animal before. The panda has been chosen as the universal symbol of endangered animals. The only natural habitat of pandas is in China, deep in the bamboo forests, at an altitude of 5,000-10,000 feet. It is a remote relative of the bear, but unlike the bear, the giant panda is a vegetarian. It is very fond of bamboo shoots: it eats 20 to 30 pounds of shoots each day. If it feels threatened by a brown bear or a panther, it climbs into a tree and stays there until the danger has passed. Its front paws are distinctively endowed with six fingers, which help the panda more easily grasp the bamboo stems. The panda is a solitary animal; it joins its kind only during the reproduction period. After a five-month gestation period, the female builds a litter, where she gives birth to one cub the size and weight of a mouse. The cub suckles for six months, and the mother watches over it until it is three. As an adult, the panda stands about 60 inches tall and weighs more than 330 pounds.

THE BROWN BEAR

on't let the memory of your teddy bear fool you; the bear is an unpredictable animal, and its vigorous paws are armed with strong claws that make the bear a fearsome enemy. The bear generally feeds exclusively on plants, mushrooms, and insects. It adores honey, of course, and from time to time, it likes to catch fish. When winter is near, the bear builds a comfortable den for its long period of hibernation. Starting with the first snows until the end of winter, its life slows down; even its heart rate and respiration slow down. However, it can still watch out for possible dangers outside. Besides, during the hibernation period, the female gives birth in her den to one to three cubs each weighing no more than a grapefruit. The female is a very careful and severe mother; she doesn't hesitate to kick the cubs when they stray too far from her.

THE PEREGRINE FALCON

The peregrine falcon is one of the fastest birds; according to some scientists, it can dive at the speed of almost 186 miles per hour. It is a bird of prey and feeds especially on other birds (crows, thrushes, pigeons) that it hunts on the fly, early in the morning or in the evening. The peregrine falcon is a very efficient hunter; it can spot a pigeon on the fly from more than five miles away. Man tames and trains the peregrine falcon to use this ability for hunting. The male and the female are almost identical, yet the male is smaller than the female. Peregrine falcons mate for life. The female digs a little hollow in a cliff and lays three to five eggs that she will incubate for one month. For his part, the male watches over the territory; he also nourishes the female and later the chicks, too. The chicks leave the nest at five weeks, but they depend on their parents for two more months. If they survive the first year (in spite of cold and their main enemy, the owl), then they might hope to live for 25 years.

104

In Polar Regions

THE EARED SEAL

The eared seal has bigger and thinner limbs (called flippers) than its relative, the seal. In addition, the eared seal can lift its body off the ground with its front flippers, being more agile when moving. It is actually so agile that you can often see it in sea parks, where it performs equilibrium and juggling tricks while barking happily. Just like the seal, the eared seal is an excellent swimmer. It is very fast and hunts fish and squid. You can distinguish between the seal and the eared seal not only because of the latter's agility but also because of its tiny ears. During the reproduction period, the male defends its territory on the beach, where he meets his harem consisting of up to 30 females. The female is the one who chooses the male. They mate in water, and after almost 10 months of gestation, the female gives birth to one calf on the ground. The calf suckles for two years. The mother and the calf are able to recognize one another by voice and smell. For this reason, humans must never caress young eared seals, because mere contact with a human hand might cause a mother to reject the young seal and thereby condemn it to death. There are several species of eared seals: the sea bear, the California eared seal, and the great sea lion (or Steller's eared seal), which you can see at the bottom of these two pages.

THE POLAR FOX

The polar fox has white fur in winter and brownish fur in summer, both of which North American hunters covet. The polar fox uses its long, thick tail either as a collar during bitter frosts, or as a coverlet to sleep on. It has a remarkable sense of smell; it is a fast runner and a good swimmer. By the end of the polar winter, in March or April, mating pairs begin to form, and they start preparing a den for the coming cubs. Generation after generation of polar foxes use these dens that may be 300 years old. The polar fox can give birth to an average of 11 cubs at a time; sometimes this number can extend to 22, which is a record number in all the Canidae family. The male hunts and brings food (small rodents, birds, berries, and carcasses), while the female looks after the cubs. The couple stays together for four months, until the cubs leave the den. In addition to humans, the main enemies of polar foxes are the wolf, the lynx, and the polar bear.

THE REINDEER

The reindeer lives in troops on the European tundra and on the plains of the American North, where it is called a "caribou." It moves constantly from summer to winter pastures. Reindeer belongs to the Cervidae family, like stags and elks, but is the only member of this family in which both the male and female have horns. The horns are very useful in winter, because they enable the reindeer to remove the snow from the lichens it feeds on. The reindeer is domesticated in Lapland. The Lapps use it as a draft animal, to drag sleighs (like Santa Claus's sleigh!), but its meat and fur are highly valued, too. The wolf, which attacks the weak and the old, is its main enemy. The call uttered by the reindeer is called a "snore." During the reproduction period, males confront one another by beating their heads together, and the most powerful takes over the leadership of the females' group. After six to eight months, the female gives birth to one cub, which will suckle milk until it is six months old.

THE POLAR BEAR

The white bear is the largest terrestrial carnivore; it can be 140 inches long and can weigh 1,300 pounds. Its only natural habitat is in the Arctic (at the North Pole). Endowed with very good smell, eyesight, hearing, and white fur that blends in perfectly with the snow, this bear is a fearsome hunter. It kills its prey with incredibly powerful paw kicks, and once the meal is finished, the white bear carefully hides what remains under the snow in order to cover all traces. The white bear travels following the migrations of its favorite prey: the seal. It can even live an entire summer on a drifting iceberg if many seals live there. It is a very good swimmer; it can easily swim six miles. The polar bear does not really hibernate. The male hunts during the entire winter, while the female shelters in an ice crevasse, where she gives birth to one or two cubs, which are only 10 inches long and weigh only 2.2 pounds. The polar bear has no enemy other than man, who hunts it especially for its splendid fur.

THE SEAL

The seal is a very prudent sea mammal; like all mammals, it has warm blood and suckles its cubs. It can hardly move on the ground, but once in the water, it is an excellent swimmer. It likes to stand in water, as you can see in the inset picture. Imagine, it can stay under the water for half an hour without breathing, and it can dive 330 feet to catch the fish and squid it feeds on. The thick fat layer under its skin allows it to resist the biting cold of the polar seas. Seals gather in large colonies on shores; the colonies may number several hundreds or thousands of individuals. During the mating period, the males try to captivate the females by capering around. The seal is polygamous, which means that a male mates with several females. Mating takes place in the water. The seal pup is a big ball of fur, as you can see in the picture on top of the page. Unless hunters kill it for its beautiful fur, the seal can live about 20 years. To communicate, the seal utters all sorts of growls and sharp yips. Its predators are the killer whale, shark, polar bear, and man. (The Inuit are very fond of its meat.) There are several species of seals. You can see the common seal (or the sea lion) on the bottom of this page, the sea elephant on top of the next page, and the Greenland seal, with its pup, on the bottom of the next page. The sea elephant (with a trunk-shaped nose) is the largest seal; the male can reach a length of 20 feet and can weigh more than 6,600 pounds.

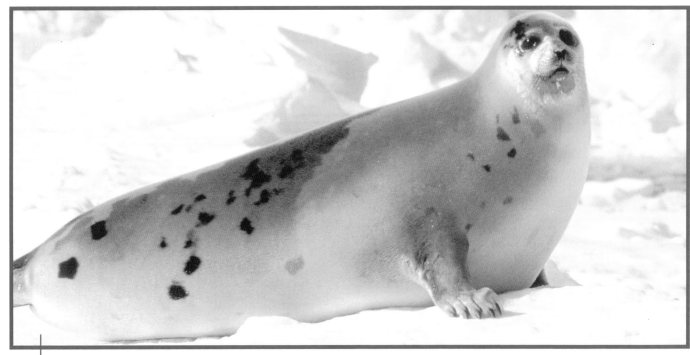

THE BLACK GUILLEMOT

Related to the razor-billed auk, the black guillemot stands almost vertically on the ground. It lives in small groups in the seas near the North Pole until the moment it must lay its eggs, when it returns to shore, where the male-female pair isolate themselves. The female lays one egg at a time; both parents incubate the egg in turn for one month. When not incubating, the male or the female goes fishing in the sea. The black guillemot catches fish under the water by propelling itself with its robust wings. These enable the bird to "fly" under the water.

After the egg hatches, the parents stay with their chick in the daytime. They leave at night to catch fish to feed it in turn. This technique ensures the chick has little exposure to predators (the peregrine falcon), and that birds like jays, ravens, or crows cannot come in the parents' absence to devastate the nest.

THE RAZOR-BILLED AUK

This bird with black-and-white plumage stands upright like an "i" and closely resembles the penguin. However, unlike the penguin, the razor-billed auk is a flying bird. Its natural habitat is the Arctic (the North Pole), and not Antarctica (the South Pole), where penguins live. It feeds on plankton, fish, mollusks, and shellfish that it hunts while swimming underwater.

The razor-billed auk lives out at sea almost the entire year; it comes back to shore only during the reproduction period. Then, it nests in colonies and prefers the craggy cliffs to protect itself against its main enemy, the polar fox. It also must pay attention to gulls, which like to eat its eggs. The female lays just one egg directly on the ground, sheltered by a few rocks.

After one month, the chick hatches, and its parents feed it with little fish. At three weeks, the young chick leaves the shelter and starts for the sea.

THE PENGUIN

The penguin lives in colonies of several thousand. They are often seen huddled together to better endure the biting cold (-58° F) reigning in Antarctica. The penguin's waterproof plumage and the thick layer of fat under its skin ensure effective protection against frost. It is a very strange bird; it appears to be wearing a tuxedo all the time. Unlike the razor-billed auk, the penguin cannot fly, but it is an excellent swimmer. Under the water, the penguin uses its wings as flippers and its small legs as a rudder. It can dive more than 150 feet down into the water to catch the fish, squid and shrimps it feeds on. In King penguins, the father is extremely devoted; he is the one that incubates the egg for two months. A few days after hatching, a group of adults gather and look after the young just like in a nursery. It is very easy to recognize the King penguin by the vivid yellow spots on its head, neck, and beak. The King penguin stands up to 48 inches tall, thus being the tallest penguin of all. In the little Adelie penguins (they measure 18 inches), both parents incubate the egg before taking their young to the "nursery."

THE KILLER WHALE

The killer whale (or orca) is not a fish, but a mammal related to the whale and the dolphin. It is actually the fastest sea mammal, being able to reach the speed of 37 miles per hour. The killer whale can be 33 feet long. It generally lives in family pods, sometimes in communities consisting of several families. It is a ferocious predator, deserving its name of killer whale. It hunts all sorts of prey: fish, seals, walruses, squid, and penguins. It swallows them without chewing first. The killer whale even attacks dolphins and younger whales. You can often see a killer whale spreading its 22,000 pounds of muscle and fat on an ice floe, trying to eat its inhabitants (seals or penguins). At birth, the killer whale calf already weighs 440 pounds and may reach the length of 100 inches. The mother suckles the calf, and it stays with her for several years. Killer whales communicate by very sharp sounds (ultrasounds) that constitute a kind of song.

THE WALRUS

The walrus is an amazing sea mammal living in the Arctic Ocean. It looks like a huge seal with very wrinkled skin. It has a pair of big ivory teeth (they are superior canines) and male walruses can be 3.3 feet long. The walrus uses its teeth like spades, to find seashells and crabs in the sandy bottom of the sea. It swallows them whole to be ground up later in its stomach. The teeth also serve as weapons during the mating period, when males fight for the females. The walrus moves slowly on the ground and spends most of its time lazing on pebble beaches. But, once it reaches the water, the walrus is a first-class swimmer. Polar bears, killer whales, and the Inuit are its main enemies. The Inuit consume walrus meat, use its skin to build their canoes, and manufacture weapons, tools, and art objects from its bones and ivory teeth.

THE MUSK OX

In terms of fur, the musk ox holds the record for longest among mammals because of its long-stranded fur. The strands of its fur may be 3.3 feet long and form a thick woolen coat, allowing it to live very well at the glacial temperatures of the regions where it lives, namely the Canadian arctic tundra, Alaska, and Greenland. These temperatures may go down to nearly -68° F. The musk ox lives in herds of up to 60. During their long mating dances, males utter deep bellowing sounds and fight violently to win the females. The female gives birth to one calf every two years. When wolves attack the herd, the adults gather together, circling the calves and using their sharp horns to defend themselves.

In Seas and on Seashores

THE WHALE

ike all mammals, the whale suckles its young and breathes with its lungs (at the surface of the water through nostrils situated on the top of its head). The blue whale is the largest animal in the world. At birth, it weighs 4,400 pounds and is 23 feet long; as an adult, it can reach the length of 100 feet and weigh 330,000 pounds (as much as 25 elephants!). Despite its weight, the whale is a very good swimmer; it can reach the speed of 31 miles per hour and can jump out of the water (just like the humpback whale in this splendid picture). It can stay under the water for two hours and can dive to a depth of 1,600 feet. It feeds on tiny shrimp (krill) by filtering enormous quantities of water through the bones of its upper jaw. Intensive fishing has almost exterminated the whales. Whale fishing is prohibited today, but, unfortunately, many countries still practice it. Except for man, the only enemy of the whale is the killer whale, which attacks young whales.

THE BARRACUDA

The barracuda is a terrifying fish that can reach the length of 100 inches. Its slender, silver body is really designed for speed. It is an extremely voracious predator that hunts everything randomly, alone or in schools. As soon as it notices a school of fish, it dashes and swallows greedily everything that passes before its sharp teeth. Usually, the barracuda swims out at sea, but it doesn't hesitate to come close to the beach. Swimmers have even been bitten in 12-inch-deep water! People fish for barracudas. But you should be very careful, because eating it is risky in tropical regions. Its meat can contain the poison produced by a microscopic alga living in the coral reefs. If people ingest this poison, they may come down with the disease known in those regions as "ciguatera."

THE CORAL

Coral can be found alone or it can form a colony (in this case, there are thousands of individuals stuck together). Each individual of the colony is called a "polyp" and looks like a sea anemone in miniature. It has a crown of tentacles with which it captures the tiny shellfish it feeds on. The polyp produces a calcareous skeleton that protects it. The skeleton of the colony may have various forms: branch, tray, or sphere. The coral is found in all seas, but it creates reefs only in warm seas. Reefs form by the accumulation of coral skeletons. This takes place very slowly; a 130-foot-high reef may be 2,000 years old! However, there are enormous reefs in the seas; the largest reef, known as the Great Barrier Reef of Australia, is 1,491 miles long! Reefs represent an incredible treasure: they attract crowds of organisms that shelter, breed, and feed there.

THE DOLPHIN

The intelligence of the dolphin is often compared with that of the chimpanzee. It utters sounds (whistles or clicks) almost continuously in order to communicate with its fellow creatures and to produce a phenomenon called "echolocation." When these sounds meet an obstacle, they bounce back producing an echo that enables the dolphin to orient itself, avoid obstacles, and spot its prey (squid, shrimps, fish). The killer whale, its close relative, is its main enemy. The dolphin can stay under water for three to four minutes. When sleeping, it floats at a depth of 20 inches and comes to the surface every 30 seconds to breathe without even waking up. The dolphin is a sociable animal that lives in pods of several hundreds or thousands, which have no leader. Sometimes you can see a dolphin coming to the aid of another wounded or sick dolphin or even a man.

THE STARFISH

The starfish is related to the sea urchin. It lives in all the seas of the world. Its diameter is between two and 40 inches, and as you can see in the pictures, it comes in various shapes and colors. It generally has five limbs, but some starfish have 50! If one of them is severed, it can easily regrow. Thousands of minuscule "legs" are found on each limb. Each limb is provided with suckers, enabling the starfish to move. Some starfish feed on mollusks, others on sponges or corals, or even on other starfish. They have a very peculiar method of feeding; they pull their stomach through their mouth and digest the prey outside their bodies. The starfish has few enemies. In fact, its skin contains substances that give it a completely unpleasant soap taste. Some starfish are even covered with venomous spines.

THE SEA HORSE

The sea horse cannot bite, so it must suck up the little shellfish and tiny fish it feeds on. Its eyes move independently from one another, enabling it to perceive prey in various directions without moving its body. It is the only fish that swims in a vertical position. It is propelled by two little fins on its back. They move very rapidly, but they are not very efficient. Therefore, the seahorse is not a very good swimmer and is easily disturbed by the slightest current. For this reason, it often hangs on to sea plants or algae, winding its tail around them. Seahorses come together during the mating period. The male and the female link their tails, and the female introduces 200 "eggs" into a pocket on the male's abdomen (similar to a marsupial's pouch). Therefore, with sea horses, it is the father who "gives birth."

THE MANATEE

The manatee is a vegetarian sea mammal; it can weigh over 1,300 pounds and reaches a length of 16.5 feet. It lives mainly in rivers and estuaries along the Atlantic coasts of Africa and Latin America, in small groups consisting of a dozen individuals that spend most of their time at the bottom of the water, grazing on plants and roots. For this reason, it is nicknamed the "sea cow," which fits better than the name of "mermaid." However, mermaid legends originate with the manatee and the dugong, its relative from the Indian Ocean. The manatee has smooth skin like other mammals in the Cetacea family. Although it looks a little like a big seal, don't be fooled by its appearance. It belongs to the ungulate family, together with the seahorse, having adapted to an aquatic environment. The manatees are very peaceful animals; they never fight, and they mate for life. Every three to five years, the female gives birth to one baby that stays with her almost two years.

THE JELLYFISH

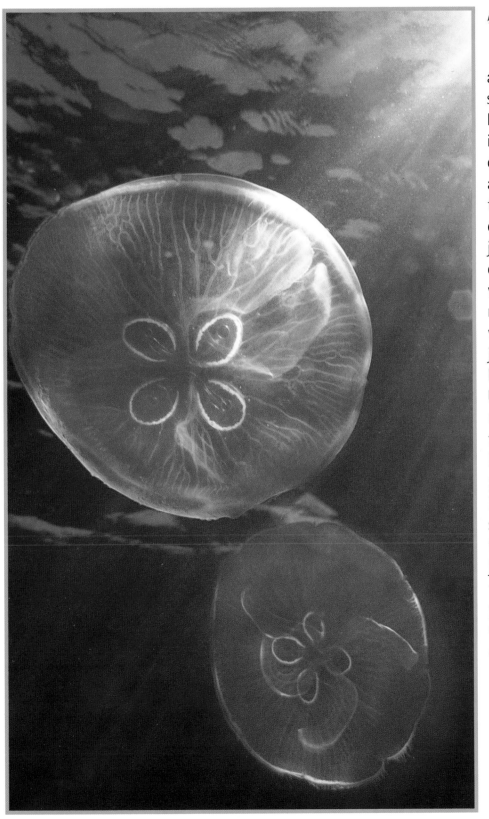

The jellyfish are related to corals and sea anemones, but they are not fixed; they float on sea currents. Once in their lifetime, some jellyfish change into fixed polyps and later on change back into jellyfish again. There are countless types of jellyfish, in a variety of shapes and colors. The jellyfish is generally formed of a disk (called an "umbrella"), with a mouth in the middle, many tentacles, and a lot of water. For this reason, the jellyfish "melts" in a few hours when it washes up on a beach. The giant jellyfish has more than 1,000 smooth tentacles that can reach the length of 16 inches. They are poisonous and represent a huge killing net, where shrimps and little fish can be trapped. Yet, the largest jellyfish are not necessarily the most dangerous. The most poisonous jellyfish live in warm seas, and their sting can be lethal to humans.

THE MORAY EEL

Although it looks like a snake, the moray eel is a fish; it belongs to the same family as the conger eel. The longest (about 138 inches) and the most colorful moray eels are found in coral reefs. It is a solitary, nocturnal fish; when night falls, it takes after its favorite diet: shellfish, octopuses, and other fish. In the daytime, it hides under the rocks or in a reef hole. You can see it in the pictures here, taking its favorite position; its body is hidden, its head pokes out of a shelter, and its mouth is open. Its sharp teeth are visible when its mouth is open, giving the moray a very impressive look. It can wound divers very badly if disturbed or if it mistakes their fingers for octopus tentacles. Fortunately, the moray eel is not aggressive (it is rather shy), and such accidents occur quite rarely.

THE OCTOPUS

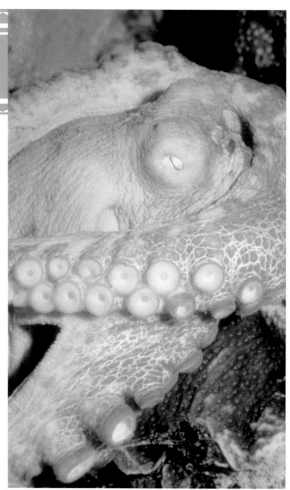

The octopus is a cephalopod mollusk of dramatically different sizes, between 1.2 and 355 inches long, like the Pacific giant octopus. It has two big, globular eyes in its head, from which stream eight long tentacles provided with suckers. Instead of a mouth, the octopus has a strong "beak" that you must beware of; it can be poisonous in some tropical octopuses, and its sting may be lethal. The octopus is the most intelligent of the invertebrates; it can unscrew the lid of a jar to reach its prey. The octopus moves by clinging to rocks with its tentacles, which it also uses for catching prey (shellfish, mollusks, tiny fish); the octopus locates its prey by sight, touch, or taste. When threatened, the octopus produces a cloud of ink and takes advantage of the sudden "dark" to run away; it can also change the color of its skin to camouflage itself in less than a second.

THE PORCUPINE PUFFER

This tropical fish sometimes can be longer than 20 inches. It is covered with spines that normally run along its entire body, but it can inflate at the tiniest menace by filling its stomach with water or air. It can double or even triple its size to look like a big, round sack covered with spines. When inflated, it looks like a pincushion, or porcupine, hence its appropriate nickname. No one can catch hold of it without getting injured, and this is why its enemies are discouraged. People used to dry it in this shape to manufacture souvenirs, lanterns, and lampshades. The porcupine puffer feeds on oysters, corals, sea urchins, and crustaceans, breaking their skeletons easily with its strong, sharp "beak." This beak is actually made of its only two teeth (one in each jaw), hence its Latin name of Diodin, which means "the two-toothed fish."

THE CLOWNFISH

The clownfish (a type of anemone fish) is a small fish, between 2.4 and six inches long, which is part of the coral reef landscape. Characteristically, this fish lives in the neighborhood of a large sea anemone. When it detects the slightest danger, it immediately takes refuge within the tentacles of the anemone. In fact, the painful sting of the burning tentacles of the anemone can sometimes be lethal to all animals except the clownfish. Thus, the anemone offers it the perfect refuge-and the anemone can feast on the leavings of the clownfish's meals. Clownfish live in pairs, and they fiercely defend their territory. Once the pair finds "its" anemone, it seldom leaves it thereafter. During the reproduction period, the male cleans part of a stone in the neighborhood of the anemone, where the female lays between 300 and 700 eggs. The male will look after the eggs till they hatch.

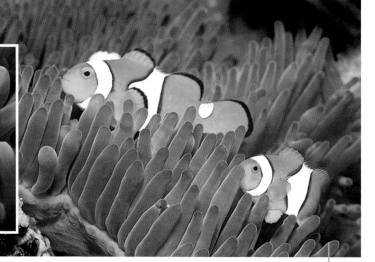

THE RAY

Rays are fish whose skeleton is not made of bones, but of cartilage. They are related to sharks. The ray swims very gracefully by way of the swaying movements of its fins (also called "wings"). It generally feeds on the oysters, crustaceans, and tiny fish it catches when searching through the sand. Some of them feed on tiny shrimps (zooplankton), by filtering large quantities of water, in a way similar to that of whales. This is true of the sea bat, also called the sea devil (pictured in the inset photo below). The sea bat is the largest ray; its wingspan can reach 26 feet. The long, thin tail of the sea bat has one or two poisonous spines, which are capable of causing serious injuries to its enemies. There is also a species of electric rays. They are real, natural electric batteries capable of producing an electric discharge of from 12 to 200 volts in order to stun or kill their prey or enemies.

THE LEATHERBACK TURTLE

Many turtles live in water; their carapace doesn't bother them that much in water, being easier to maneuver here than on the ground. Therefore, it is obvious that the biggest turtles live in the sea. The leatherback turtle is known as the biggest turtle, because it can weigh 1,700 pounds and can be 100 inches long. It lives out at sea, feeding on algae and fish. Like all other sea turtles, the leatherback turtle must go on shore to lay its eggs. Perhaps you have already seen, in a TV documentary, a giant green turtle carrying its 440 pounds with difficulty onto a beach to dig a hole for its eggs. After the eggs hatch, the young turtles start toward the sea alone. If they get there, they have a chance to live 120 years. Some turtles (for instance the green turtle) are still hunted by man, who cooks traditional turtle soup of its meat.

THE SHARK

ust like the ray, the shark is a cartilaginous fish; its skeleton is made of cartilage rather than bone. There are several types of sharks. The smallest shark is 28 inches long (the spotted dogfish); the longest sharks, like the whale shark and the basking shark, reach more than 50 feet. Fortunately, these huge sharks are harmless: they feed only on tiny shrimps. However, other sharks, like the tiger-shark and the leopard-shark, are extremely dangerous; both fish and humans are afraid of them. The famous great white shark reaches 40 feet in length and can weigh 6,600 pounds. It is one of the most dangerous sharks. It has been classified as a "man eater," and this nickname suggests the harm it is capable of inflicting. Hundreds of razor-sharp teeth are found in its jaws. In all sharks, the teeth, which periodically are shed, are arranged in rows; sharks replace their teeth all their life.

The bald eagle is probably one of the most famous eagles. It is the emblem of the United States of America and is represented on the quarter and half dollar coins. The bald eagle may have a wingspan of more than 6.5 feet. It lives near seas and lakes on almost all continents. It feeds mainly on fish that it catches on the surface of the water. On some occasions it can attack birds and small mammals. Its sharp claws are extremely strong, and sometimes the eagle cannot turn loose of its prey, and if that prey is very heavy, it can draw the eagle under the water and drown it. Bald eagles prefer nesting in tall trees and use the same nest year after year. The female lays two eggs, which she incubates for 35 days. The chicks take flight at the age of two months. They are entirely brownish and can be mistaken for royal eagles. As you can see in these pictures, the head and tail of adult bald eagles are completely white.

THE BALD EAGLE

THE CRAB

Crabs come in different sizes and shapes, depending on the species; common characteristics of the crab include its protective, thick shell and its five pairs of legs, of which the front pair have been transformed into pincers. These pincers are huge sometimes, like in rock crabs (or edible crabs). In other species, such as the velvet swimming crab, the hind pair of legs have developed into fins, allowing the crab to swim. The crab walks laterally. It carries its eggs in a bunch under its stomach. The eggs stay there until they hatch. At the beginning, they are embryos; then they transform little by little into swimming larvae and eventually into small crabs. Once attaining this final shape, crabs must change their shell regularly to fit in them (this process is called "metamorphosis"). At this point, they are very vulnerable; to avoid predators, they must hide until the new shell is formed. Crabs feed on every kind of food they find nearby: oysters, dead fish, fish eggs, and worms.

THE CORMORANT

There are 37 species of cormorants in the world. Like the majority of sea birds, the cormorant is rather "clumsy" on the ground. It feeds on fish (that is why it is called a "fish-eater"). It is an excellent fisher. People in Asia use it like a fishing rod. They tie a string around its neck and turn it loose to find fish under the water. The string is tight enough to prevent the cormorant from swallowing the fish. Cormorants often fish in groups, and this method proves very effective when they target schools of fish. It can dive 115 feet deep into the water and use its wings and webbed feet to propel itself. After fishing, it returns to the shore to dry its wings in the sun, maintaining its famous position: wings spread wide and body perfectly still. During the mating period, the cormorant nests in large colonies on cliffs. It builds its nest on the ground, and the female lays three or four eggs.

THE NORTHERN GANNET

The Northern gannet is a large bird whose eyes are accented with gray. It is very fast when flying, but it is rather clumsy on the ground. It nests in large colonies on the shore. You can see it especially on ledges and cliffs, where it takes refuge against such predators as the fox. Its nest is made of sea plants, feathers, and soil. The female lays one whitish egg, which she incubates for 44 days. Both parents feed the chick for about three months. Feathers start to replace down at six weeks of age: at the beginning they are brownish, then they lighten progressively. The Northern gannet feeds on fish like herring and mackerel. Its fishing technique is spectacular. It flies over the water. As soon as it sees its prey, it comes down in a nosedive, sometimes from more than 130 feet high, its wings half folded. It passes the prey in the water; then using its webbed feet, it swims to the surface, grabbing the prey on its way up. The Northern gannet eats the prey immediately after reaching the surface.

THE GULL

Its habit of eating every kind of food (including the contents of trash cans) makes the gull an unattractive bird. However, it is a useful garbage collector. In fact, it cleans the beaches and the harbors of waste (dead animals, picnic trash). The large number of gulls represents a real problem. They live almost everywhere in the world. They nest in colonies on shores, in fields, and even in cities. Their nest is made of plants and trash placed directly on the ground. The female lays three olive-green eggs and incubates them for three or four weeks. The orange spot on the beak of adult gulls serves as a "target" for the chicks. When the chick is hungry, it kicks this spot on its parents' beaks; they then regurgitate food for the chick. The young gull can fly after six weeks. The silver gull chicks you can see in these pictures are brownish when growing, then become white with gray wings. The adult silver gull has black wings. Gulls mate at three years of age and can live 30 years.

THE TERN

erns are similar in shape to gulls; however, the tern has a red beak with a black tip and a black "skull cap" on its head. The tern's "ragagaga" call sounds like a giggle and generally announces bad weather. The tern is found in the south Pacific Ocean all the way to the Indo-Malaysia region. It is a coastal bird, rarely found more than two kilometers from the coast. It also lives along lakes and rivers. It feeds on little fish, crustaceans, insects, and worms. It flies over the water to fish,

and as soon as it sees its prey, it comes down in a nosedive from more than 10 feet high. Terns nest in colonies on beaches and shores, in small hollows dug in the ground and covered with leaves and twigs. The female lays two or three sand-yellow eggs with brown spots. Both parents incubate the eggs for three weeks. After hatching, the chicks are fed by the parents for two months.

THE ATLANTIC PUFFIN

This cute little bird, looking quite sad, averages 12 inches tall. Because of its characteristic colorful beak, the Atlantic puffin has also been called "the sea parrot." It moves prudently on the ground by jumping; it is not very skillful at flying, and it often tumbles down when landing. Nonetheless, the Atlantic puffin is a first-class swimmer. It chases its prey under the water and returns with the captured wriggling fish in its beak. It spends the winter out at sea. It returns to shore during the mating season and nests in large colonies. The Atlantic puffin shown in this picture nests on the northern coasts of the Atlantic. Atlantic puffins mate for life. The female lays only one egg, and both parents incubate it. After hatching, the chick is fed with small fish for two to nine months. After this period, the parents return to the sea. The chick must leave the nest on its own to learn to dive and fish.

THE PELICAN

The pelican is one of the largest sea birds and one of the most amazing. Its long, flat beak and the extensive crop under that beak give the pelican a unique look. After catching a fish, the pelican stores it in its crop like in a pantry. Sometimes the crop is so full that it overflows. When two pelicans fraternize, they inflate their crops, lift their beaks vertically, and bend their heads slowly. During the mating ritual, the male executes a dance around the female by beating on the ground with his webbed feet. Pelicans build their nests directly on the ground, on an island away from predators. The female lays between two and five eggs, and both parents incubate them in turn for one month. The young pelicans start flying at the age of three months. Until then, they feed on half-digested food from their parents' crops.

THE OYSTERCATCHER

The oystercatcher belongs to the same family as the sandpiper. It is a little, long-legged bird, standing up to 15 inches tall; it is very easy to recognize the oystercatcher because of its red beak and its black and white plumage that makes it look a bit like a magpie. It lives in the coastal regions of Europe and Africa. It swims and dives very well and is found most often on beaches and in lagoons. Its name derives from its diet of oysters. It uses its sharp beak to open the oysters. The female lays between two and four eggs, and she incubates them for a period that varies from 24 to 27 days. The chicks leave the nest at 40 days old, when they are able to fly.

THE SANDPIPER

The sandpiper is related to the curlew and the oystercatcher. Depending on the species, its size varies from that of a skylark to that of a starling. The sandpiper loves the swampy regions of seashores, where it lives in large flocks. It searches the silt with its long, straight beak, looking for tiny invertebrates that it feeds on (oysters, worms). Some species are monogamous (they have just one mate), others are polygamous (each male has several females); others, like the shore sandpiper shown in this picture, are polyandrous (which means that each female has several mates). The female lays between two and five eggs that must be incubated for 18 to 32 days. The chicks leave the nest very soon after hatching.

THE HERMIT CRAB

The hermit crab is the only crustacean with no carapace to protect its abdomen. It is therefore highly exposed to predators (other crustaceans, fish, sea birds). However, it has found a solution, which is dwelling in empty snail shells. It hangs onto the shell with its long abdomen; its head and the legs generally remain outside the shell. It leaves its shelter only to find a larger one when sloughing or mating. You may have the chance to see a hermit crab outside its shell. It is an asymmetric creature; it looks like a lobster with a twisted body. It has five pairs of legs; the first pair has changed into pincers. Even the pincers are asymmetric: one of them is much bigger than the other. The hermit crab feeds on mollusks (oysters) and dead animals.

THE SPINY LOBSTER

reat lovers of seafood value the spiny lobster highly. This crustacean is a close relative of the crab and the lobster. Its antennae are very long and serve as "radar." During their spectacular migrations, you can see thousands of spiny lobsters walking in line one after the other. The spiny lobster generally lives at depths of 230 to 660 feet in the sea, but it comes closer to the coasts during the breeding period. The male turns the female over and lays its spermatozoa on the female's

abdomen. A few hours later, the female ovulates and is inseminated. The inseminated eggs are crowded together in gelatin and remain stuck on the female's abdomen. After 20 days, a swimming larva emerges from each egg. After several sloughings, the larva will transform into a little lobster. The lobster grows during its entire life just like all crustaceans: it grows progressively with each sloughing.

THE SEA URCHIN

The sea urchin is related to the starfish. This echinoderm generally has a spherical body, but in the case of sand sea urchins, the body can be flattened. The body of sea urchins is covered with spines that can be from 0.2 inches to 12 inches long. Depending on the species, sea urchins can be herbivorous (feeding on algae) or sand eating. There are also carrion-feeding sea urchins (feeding on dead animals). The mouth of the sea urchin is found on the underside of its body; herbivorous sea urchins have five teeth forming what we call "Aristotle's lantern," which enable them to "graze" on algae. Sea urchins breed in water: the female lays eggs that transform into embryos and then into larvae, if previously inseminated by the spermatozoa of males. The larvae swim for a few weeks, then they fix on the bottom of the sea or on an alga, and they transform into little sea urchins. People highly value certain sea urchins, which represent a dish as elegant as caviar.

THE SHRIMP

You have surely fished for these little crustaceans with a shrimp net at least once in your life. Depending on the species, shrimps may be from 0.2 up to eight inches long. They are related to crabs and lobsters, but unlike their relatives, shrimps do not have big pincers. They can be transparent or colorful. Their colors vary from gray to light red and from blue to yellow. Regardless of their initial color, once they are cooked, all shrimps end up pink. When peeling a shrimp, you sometimes may find little red spheres on its abdomen; these are the eggs (more than 1,500 sometimes). After hatching, little larvae emerge from them. By the end of a few weeks, these larvae transform into adults. Shrimps mainly feed on the leavings of other sealife and dead animals. They also feed on worms, tiny crustaceans, and plants. But they are a good prey for countless fish and marine mammals as well. In fact, whales feed on krill, which is a species of small shrimps.

On Lake-Shores and River-Banks

THE ANACONDA

Anacondas belong to the same snake family as boas and pythons. They are one of the largest snakes in the world, reaching a length of 30 feet (even 36 feet, according to some people), and they can weigh more than 220 pounds. The Amazonian Indians call anacondas "the spirit of woods"; they respect and fear them. The anaconda lives in South America, always near a stream. It is an excellent swimmer (it swims four times faster that a swimming champion) and spends much time in water. It often coils around the branch of a tree, half hanging in the water, and slips into the water when it is disturbed or when some prey passes by (rodents, birds, tortoises, and young caimans). It kills its prey by suffocation and can swallow the equivalent of its weight at a single meal. The adult anaconda has no enemy, not even the powerful caiman. The female gives birth to about 30 little snakes (already formed), approximately 24 inches in length, which grow 30 feet every year until sexual maturity at about the age of four.

THE BEAVER

The beaver is one of the largest rodents: it can be more than 3.3 feet long (including its tail) and can weigh more than 66 pounds. Its teeth are strong and allow it to play "lumberjack." The beaver needs only five minutes to cut down a tree four inches in diameter. It uses the tree trunks to build and strengthen a dam, which can be 900 feet long and 10 feet thick. The dam serves to keep the water level high where the beaver builds its dwelling out of branches arranged to form a kind of hut, with an exit under the water.

Beavers are vegetarians and their favorite diet is bark. Once a pair mates, the beavers stay together all their lives. After a three-month gestation period, the female gives birth to two to five pups. Today, beavers are protected by law, but they have been hunted for a long time by North American trappers for their highly valued fur.

THE TOAD

The toad belongs to the batrachians. Its skin is often rough and is spotted with warts behind the eyes and on other parts of the body. These warts indicate the places where its poison glands are found. But don't worry, you can take it in your hand without fear; except for some tropical species, the venom of the common toad is not poisonous to humans. Unlike other frogs, the toad lives a good deal of the time away from water sources. Yet, it breeds in water, and therefore it must often migrate long distances to find a breeding place (sometimes several kilometers). The male calls the female by echoing her croaking; when she comes, the male jumps on her back. After a while, the female lays between 5,000 and 20,000 eggs that the male inseminates in the water. These eggs are crowded together in a kind of gelatin, and sometimes they form a 13-foot-long string. One month later, tadpoles emerge from the eggs. In three or four months, they change into toads and get out on the ground. Like all other frogs, the toad has a long tongue that unfolds when it's catching a moving insect. (The toad can see only moving objects.) Toads also feed on worms and slugs.

THE ALLIGATOR

This reptile is found only in the southeast regions of the United States and in China. Like the crocodile, its relative, the alligator is at ease both in water and on the ground. It hunts at night. It is a rather lazy hunter; generally it finds a place under water and waits for prey to pass by. It can stay under water more than one hour. It also can catch birds or large mammals that come to drink water right where it hides. During the mating period, the male defends its territory and mates with several females. They copulate in water, at night. Shortly after that, the female lays some 20 to 80 eggs in a hole dug in the bank and remains there to supervise them. The alligator can live 50 years and can reach a length of 20 feet. Hunting almost caused its extinction (its skin is highly valued), but it has bred again in the United States.

It remains a very rare species in China, living only in the Yangtze River.

THE CROCODILE

The crocodile is the heaviest of today's reptiles; it can reach 4,400 pounds! It is related to the alligator, yet the two can be differentiated easily: the crocodile has longer jaws, and even when its mouth is shut, two teeth are still visible in its lower jaw. The crocodile lives in various regions, including Asia, Africa, and Australia. Its length varies from 60 inches (the dwarf crocodile) to 33 feet (the marine crocodile). It is a strong predator; without hesitation, it attacks zebras and antelopes that come to drink. It also feeds on birds and fish. It likes to sunbathe and let small birds clean it of leeches. These birds also clean its gums. Every year the female lays about 30 eggs in a hole dug on a beach and covers them with sand. She stays close to the nest to watch over it. When the hatchlings are about to come out, they make noise, announcing to their mother that it is time to uncover them and help them get out of the nest. From that moment, each of them has a chance to live at least 70 years.

THE SWAN

The swan is an aquatic bird, famous for its beauty and its graceful flight. This is why it is not at all surprising that it has been the subject of countless tales, ballets, and paintings throughout the centuries. The swans pictured here are called "mute swans," or "domesticated swans." Their splendid plumage consists of nearly 25,000 feathers. But you should know that not all swans are domesticated; many of them migrate, changing their habitat depending on the seasons. Also, not all swans are completely white; some swans living in Australia are actually entirely black. Swans live on the surface of lakes and feed on aquatic plants, frogs, and small fish. They are imposing birds. They can weigh 44 pounds and can have a wingspan of more than 6.5 feet. Swans mate in spring. The male can be distinguished by a little red spot at the base of his bill. The female nests at the water's edge and lays five to eight eggs. Young swans are gray in color.

THE PINK FLAMINGO

This elegant bird is a wading bird whose 60-inch-long legs end in webbed feet. Its curved bill is very unusual: it is provided with smooth lamellae that filter water, enabling the bird to eat its diet (algae, tiny shellfish, mollusks, and worms). This is exactly what it is doing when you see it walking, its head bent and its bill in the water. Its pink color is caused by the dye contained in the little shrimps it feeds on in salty or brackish lakes. The more shrimps they eat, the pinker their plumage becomes. The male and the female copulate after a long mating dance. The female lays only one egg at a time, in a nest called a "mound" (which is actually an elevation of the earth).

Both parents incubate the egg for 30 days. At first, the chick is gray and has a flat bill. The chick is fed on regurgitated food from its parents until the age of two months, when its bill starts curving and it is able to feed itself. Its plumage starts turning pink toward the age of one.

THE GREAT EGRET

Great egrets are of the same family as herons. Their long neck is usually in an S-shaped curve; they have long bills, yellow eyes, and long legs. Smooth, long, beautiful feathers appear on the back of egrets during the mating period. At the beginning of the 20th century, these special feathers (called "egrets") became very fashionable ornaments for ladies' hats. Today, egrets are protected by law. Egrets nest on platforms made of branches, in trees or directly on the ground. The female lays three to five blue-yellow eggs. After about 25 days of incubation, the chicks emerge; the parents look after them and feed them on fish, frogs, and little reptiles. Adult egrets feed on lizards, insects, and even rabbits.

THE CRANE

Cranes are elegant, long-necked, short-billed birds; they walk very gracefully. The calls uttered by cranes sound like cries, trumpet sounds, or clattering. When they are flying, it is easy to distinguish them by their long neck and long legs giving them a continuous body line. At maturity, the crane chooses a mate for life. The couple nests in swamps or damp meadows. The nest is built directly on the ground, and the female generally lays two eggs. The parents incubate the eggs in turn for one month. Shortly after hatching, the chicks are able to follow their parents; at the age of two months, they can already fly. The crane shown in the pictures here is called "the crowned crane." Can you guess why? Because it indeed wears a real "crown" of fine gold-colored feathers on its head. Cranes feed on greenery, but also on insects, mollusks, and little vertebrates, like frogs.

THE FROG

As you can see in the photos, there are frogs of various colors and sizes. Frogs are amphibians: they live both in water and on the ground. Their webbed hind feet allow them to swim very fast, almost as fast as people when using flippers. Frogs move on the ground by walking or leaping. Some frogs, like the ones in the two pictures below, are tree-dwelling frogs and climb in trees. During the warm season, the male attracts the female by its well-known croaking, which we often hear during spring nights. Frogs mate in water; the female lays several thousand eggs, which the male immediately inseminates. Tadpoles appear a few days later. They change into frogs little by little (this process is called "metamorphosis"): the legs appear and the tail disappears. By the age of three months, their lungs have formed, and they can get out of the water. Frogs feed mostly on insects; they launch their long, sticky tongue with great precision toward the prey and catch it.

THE HERON

Herons are wading birds feeding on little fish, shellfish, insects, amphibians, and little reptiles that they catch in shallow fresh or salty water. They swallow their prey just like that. The heron shown in the picture is the "grey heron." As you can see in the picture below, its eyes are yellow. It is easily distinguished when in flight because of its S-shaped neck. Every year in spring herons return to the same spot to mate. The male then defends a place in the top of a tree where he intends to nest. He courts a female, which lays three to five pale blue-green eggs shortly after. Both parents incubate the eggs: the male minds the eggs in the daytime and the

female at night. After the little herons appear, the parents start looking for food in turn, day and night. This action lasts until the chicks are about two months old and leave their parents.

THE COOT

This bird reaches a length of 14 inches. It is often mistaken for its relative, the common moorhen. Like all coots, the one shown in these pictures has black feathers. It is the European common coot and is distinguished by its white bill and forehead. Coots are sociable birds; they gather on the surface of lakes and large ponds to winter. Imagine the deafening noise they make when simultaneously uttering their characteristic call: "kau . . . kau." Coots nest in the reeds, using the same nest every year. The female lays between four and 11 eggs, which she incubates for 21 to 24 days. The chicks learn to fly about the age of two months. Coots feed on aquatic greenery. To root them out from the bottom of the water, they can dive more than 23 feet deep and can stay under the water for 30 seconds.

THE DRAGONFLY

The dragonfly is one of the most ancient flying insects on Earth. Many million of years ago, dragonflies were the size of gulls. Nowadays they are no more than four inches long. They fly fast and silently (they can reach a maximum speed of 50 miles per hour) and can change direction suddenly, turning right or left, hovering in place, or even flying backward. Dragonflies live close to rivers or ponds, which is where they are born. After mating during a "wedding" flight, the female lays eggs on aquatic plants. A larva emerges from each egg and lives in the water from one to five years. Dragonfly larvae are very active predators: they hunt mosquito larvae and tadpoles. They slough nine times during their aquatic life. During the final sloughing, they hang onto aquatic plants and transform into adult flying dragonflies. The adults live only several weeks, but they are fearsome predators just like the larvae, feeding on flies and butterflies.

THE OTTER

The otter lives near fresh waters (lakes, streams) and salty waters as well (seas, oceans). The body of this little mammal is perfectly designed for swimming; its short feet are webbed, and its long tail serves as a rudder. The otter can also stay under water for eight minutes without breathing! It is a very efficient hunter. It feeds especially on fish, but also on frogs, birds, and mollusks. Its whiskers are very long and enable it to detect and spot its prey in the dark or in muddy waters. It digs its burrow in the bank of a river or lake. It has two exits: one exit leads directly to the water, the other to the ground. The female gives birth to two to five whelps after a nine-month gestation period. Otters communicate by whistling or chuckling. They are intelligent, very sociable, and playful. This is why it is very easy to train them, especially when they are young.

THE RACCOON

The raccoon has a very unusual appearance. Among its characteristics we can count a tail with black and white stripes, very agile front paws, real miniature hands, and, of course, the famous Zorro's mask it wears. It is a harmless, sociable, and intelligent animal. It is easy to train when young, and it can return to the wild and adapt again after having lived among people, which is quite rare in animals. The raccoon is a very good hunter; it rummages all night long near the water in search of food: corn, shrimps, frogs, fish, birds, and fruit. Did you know that it always washes food before eating it? The male is polygamous (it inseminates multiple females) and the female is monogamous (mates with only one male). The couple forms when mating. The female gives birth to three to five cubs two months later and looks after them alone. The predators of the raccoon are pumas, lynxes, coyotes and . . . man, who hunts it for fur.

THE SALAMANDER

The salamander is a batrachian just like its relatives, frogs and toads. The salamander looks a little like a lizard (which is a reptile), but differs from it by lacking scales and claws and by its always-wet skin. Whether terrestrial or aquatic, the salamander always needs humidity to live. The female lays her eggs in water and the larvae are aquatic. Both larvae and adults are carnivorous. The larvae feed on tadpoles and tiny aquatic invertebrates, and the adult salamanders hunt earthworms, slugs, spiders, and insects. The salamander can relinquish its tail, which sometimes allows it to escape from its predators. Another tail, shorter than the first, will grow after that. The salamander is generally between two and eight inches long. Some species, however, can reach a length of 20 inches. The giant salamander from China is a very amazing species: it can be more than 60 inches long and can live up to 50 years.

THE COMMON KINGFISHER

This chubby colorful bird is very charming. Its size varies from 6.3 inches to 14 inches. Its big head and long, strong bill give it its characteristic silhouette. As its very name indicates, the common kingfisher feeds by fishing for its prey, using a highly elaborate method. Standing up on a branch near the water, it scans the water with its penetrating eyesight and waits for a fish or a frog to pass. Then it dives in the water at high speed, catches the prey, and carries it away. It returns to the branch and kills the prey by striking it on the branch, and then it swallows it whole. Kingfishers live alone, except for the period when they mate and form pairs. The "lovers" dig a nest in a bank, and the female lays five to seven white eggs that are incubated by both parents for three weeks. The main predators of kingfishers are the water rat, the weasel, and the fox.

THE STORK

Storks have long, straight, red bills; they walk and fly gracefully. The common stork shown in these pictures has large, white wings with black borders; it can stand about 3.6 feet tall and lives on plains and steppes. There is also a species of black storks; rather rare, they are smaller than common white storks and live in woods. Storks feed on little rodents, amphibians, reptiles, fish, and insects. During winter, they migrate to the warm countries of Africa and return to Europe to nest when March is near. Then you can find them in countries such as Holland, Belgium, Hungary, Poland, and Romania. Storks nest in trees and on chimneys, too. The stork's nest is huge: it is 60 inches in diameter and is made of branches, soil, grass, and straw. The female lays three to five eggs, which are incubated by both parents for 45 days.

THE NORTHERN LAPWING

The northern lapwing is a 12-inch-tall bird, distinguished by its always-raised black crest. This crest is made of four to six very thin, black feathers. When in flight, it is distinguished by its round silhouette and its white and black wings. The northern lapwing originally lived in swampy regions. Today, it lives on prairies and in humid regions, too. It is a migrating bird: it lives in Europe between March and July, and then it leaves for the northwest of Africa. It feeds on insects, larvae, worms, mollusks, seeds, and greenery from time to time. It utters characteristic sounds, depending on whether it is in flight or on the ground: it utters sounds like "pic-uit" or "ki-uit" when flying and "ti-ti-ti-ti" when on the ground. It nests in a little hole in the ground covered by dry grass. The female lays four eggs here and incubates them for 25 days.

THE IBIS

This very beautiful bird you see in the picture above, is called the "red ibis." It stands between 24 and 35 inches tall and has a characteristic long, curved bill just like all ibises. Its magnificent red color comes from a dye called "carotenoid pigment" that is found in the prey it hunts. The ibis feeds mostly on crabs and other tiny aquatic shellfish as well as on insects and even little reptiles. The red ibis lives in large colonies along coastal regions and in the northern swamps of South America. The red ibis is closely related to the best-known black and white ibis, called the "sacred ibis." During the era of the pharaohs, Egyptians believed the sacred ibis was the messenger of the god Thot. During the warm season, females build nests in the trees found in swampy regions. They lay three or four eggs, and the parents take turns incubating them. The chicks emerge after 24 days.

THE CANADA GOOSE

The Canada goose is the most famous wild goose. It is also the most widespread goose in North America. The inhabitants of Quebec call it the "bustard." It is a little smaller than the domestic goose, and its head and neck are black. Canada geese fly in a flock, which has a leader. The other geese fly behind the leader in the shape of the letter "V". The arrival of these famous V groups in the Canadian sky announces a change of season: spring is coming! Canadian geese mate for life. During the breeding period, the pair nests in a bush near a river or, more often, on a little island in the middle

of a river or lake. The female lays five or six eggs that are incubated for 25 to 30 days. The fledglings start flying around the age of two months. The small family is very close knit, and the male defends it with tenacity. It is a very fearsome fighter that can withstand the attacks of a fox; the strikes of its wings are amazingly powerful.

THE CURLEW

ust like the northern lapwing, the crane, and the stork, the curlew is a migrating wading bird. It stands about 20 inches tall and is rather easily distinguished by its characteristic long, thin, curved bill. The curlew lives on swampy plains and in peat bogs. It nests directly on the ground in a little hollow covered with twigs. Both parents incubate the eggs. The chicks are able to feed themselves very soon after hatching. At the beginning of summer and after the chicks have appeared, the curlews can be found atop hay bales or poles that mark the fields. The curlew feeds on the insects, larvae, mollusks, and worms it digs out from the soft ground with its long bill. Sometimes it also feeds on shoots and berries. Its call sounds like a sort of "curlew, curlew, curlewculi," which is the origin of its name.

THE PIKE

The pike has a long, broad beak full of sharp teeth; its body can reach a length of 60 inches. It is also called the "fresh water shark" because of the way it looks. Of course, the pike is not a shark, but it is a predator that is as greedy and fearsome as the shark. The pike is solitary and stalks its prey. It stays still, hidden beneath aquatic plants; when a school of little fish inadvertently come close to its hiding place, it charges, catching its prey and swallowing them from tail to head. The pike lives in deepwater streams with rather slow currents and in lakes and warm ponds with grassy banks. The female lays 30,000 to 60,000 spawn in a shallow area that has aquatic vegetation.

THE TROUT

The trout can reach a length of 15 to 31 inches; older individuals can be 39 inches long and weigh 22 pounds. The river trout (or the brown trout) that you see in this picture lives in Europe. Its body is dotted with red and black spots, which serve for camouflage so that the trout can go unnoticed among the rocks when light is reflecting off the water. The trout lives in cold waters and especially enjoys mountain rivers. The trout feeds on larvae, insects hunted at the surface of the water, mollusks, and tiny fish. The breeding period lasts from October to January. It goes up the river just like its relative, the salmon, to lay its spawn in special places called "reproduction areas." Once reaching these places, the female digs a real nest in the gravel with its tail. She deposits her spawn here, and the male hurries to inseminate them with his spermatozoa. After hatching, the young fish go down the river; the little fish will go up the river again when sexually mature, about the age of two to four years. Fisherman highly value the trout. They fish it with a sort of "fly," manufactured by rolling some feathers or fur around a hook, which imitates an insect sitting on the surface of the water.

On Grassy Plains and Prairies

THE BISON

With their characteristic humps, bison look much more imposing than their relatives, the cows. They are vigorous and unpredictable animals; they run very fast and charge at full speed without hesitation when danger is near. The cold winters do not represent a problem for bison; their thick fur protects them, and they find food by removing the snow with their strong heads. Bison have very good hearing and can recognize smells from distances of two miles. European bison live in forests, and American bison (the buffalo), which you can see in these pictures, live on the great prairies. Unfortunately, the American bison is found only in reservations and parks today. Bison were excessively hunted 200 years ago for their wonderful fur. This uncontrolled hunting decimated enormous herds of thousands of these animals that used to migrate long distances.

THE HARE

As compared to rabbits, hares have longer ears and hind legs and have no eyelids. The length of their hind legs is the key to their speed (they can reach a running speed of 50 miles per hour). Unlike the rabbit, the hare does not dig burrows. It rests alone in a small hole called a "litter." During the mating period, hares chase and box each other with their front legs. The female (called the "doe-hare") gives birth to from one to seven babies (called "leverets"). They are born with fur and can see and are capable of grazing grass. As in all rodents, the hare's incisors lack enamel and grow continuously. This constant dental growth forces the hare to nibble all the time in order to wear down its teeth, which otherwise would grow too long and make it impossible for the hare to feed.

THE PRAIRIE DOG

The prairie dog is not actually a dog, but a small rodent about 12 inches tall; it is closely related to the squirrel and the marmot and is very commonly found from Canada to Mexico. It also has a central European relative that closely resembles it, the ground squirrel. The prairie dog emits a barklike yip, hence its name. When prairie dogs feed on greenery in groups, there is always an individual among them that remains on alert, up on its hind legs and surveying the surroundings to watch for predators (foxes, coyotes, royal eagles, and snakes). It gives the alarm at the slightest threatening danger and jumps in a burrow. These burrows are made of rooms connected with tunnels that normally are three to five meters long. Countless burrows are placed in proximity; they form a real underground "town" inhabited by thousands of prairie dogs. At the beginning of the 20th century, some of these "towns" contained up to 40 million individuals! Today the prairie dog is an endangered species.

THE EMU

The emu is the second largest living bird, immediately after its relative, the ostrich; the height of the emu averages 6.6 feet. Its wings are very small, and therefore the emu is a flightless bird. To compensate for this flaw, it is an excellent runner. It lives in mainland Australia, in woodlands, savanna, and eucalyptus forests. It moves in small groups in search of food and water. A pair forms during the mating period, and the jealous male guards the female very closely. The female lays between eight and 15 eggs in a nest built directly on the ground by the male from leaves and grass. The male is also the one that incubates the eggs for two months and looks after the chicks until they are 20 months old. It is careful to especially watch out for eagles, which are the most dangerous natural enemies of the little emus. The chicks feed on insects; the adults also feed on seeds and fruit. Its call is very awkward; it sounds like an echo, and it is said to bellow.

THE COYOTE

The coyote has a bad reputation, which is not really justified. In reality, the coyote is dangerous neither to people nor to domestic cattle. On the contrary, by feeding on little, rapidly breeding rodents and carrion, the coyote is actually very useful! Just like wolves, their relatives, coyotes howl at night. They can also bark, growl, moan, and yell. In coyotes, the female is the one that chooses the mate, and the pair forms for several years. The female shelters in a den to bear her young. While she is tending the pups, the male defends them and brings food. The pups are easy prey for wolves, pumas, and brown bears; as an adult, the coyote can easily escape its predators, being the fastest member of the dog family. (It can reach speeds of 38 miles per hour when running.) However, there are two features that disadvantage the coyote rather often: it has the bad habit of turning around when running and of sleeping very deeply.

THE SURICATE

This little animal is related to the mongoose; it is about 20 inches long (eight inches of which are its tail). It is a harmless animal and is easy to domesticate. It is a sociable animal, living in family groups of up to 30. Its burrow may consist of several galleries, or it may form a real maze dozens of feet long and with countless exits. The female bears two to four young in this burrow. The suricate is omnivorous; it feeds on insects, little lizards, and amphibians. It hunts them and jumps on them with a rapidity and extraordinary precision, biting them at the neck. Suricates are sensitive to cold, which is why you can often see them standing up on their hind legs and warming themselves in the sun as they are doing in these pictures. They also stand this way to survey their surroundings for predators (birds of prey among others). As soon as a suricate gives the alarm, all of them immediately run into the burrow.

THE KANGAROO

The kangaroo is the best-known marsupial of Australia. It lives in troops of 10 individuals headed by a mature male. It moves on its long hind legs by hopping as much as 33 feet. It uses its long tail to keep its balance when jumping, but also to lean on when resting or to beat the ground to warn the others that danger is near. It has small front feet, with claws that enable it to pick grass and fruit to eat. It also uses its claws to comb its fur and to defend itself during fights between males. The most remarkable characteristic of this animal is, of course, the presence of a marsupial sack (simply called the "pouch") on the female's abdomen. Following birth, the little joey shelters in this soft pouch. The joey is only one inch long and is not entirely formed. It grows in the pouch and stays there until the age of seven to eight months.

THE VIPER

There are countless species of vipers. They are rather easily distinguished by their broad, triangle-shaped head. They are around 28 inches long, but the largest species can reach a length of 79 inches. Vipers generally lay eggs (they are called "egg-laying," or "oviparous," animals), but there are also viviparous vipers (which means that the female bears little snakes). The viper is a very efficient hunter; it is capable of attacking so fast that you can't even see it move. It paralyzes and kills little rodents by biting them with two long, venomous fangs. Each of them is connected to a gland, which produces venom that the viper injects when biting the prey. The viper swallows the prey whole, without chewing it, and digests it slowly.

THE ARMADILLO

ook what an awkward animal! In fact, except for the pangolin, the armadillo is the only mammal that has a carapace, which covers its body from head to tail and is made of bony slates covered by horny material. Its carapace is real armor. When it feels threatened, it rolls into a ball to protect itself. The armadillo feeds on worms, greenery, and especially on termites and ants. It detects them with its excellent sense of smell and digs into the ground incredibly fast to pull them out with its sticky tongue. This little animal, which is 10 to 24 inches long, can live everywhere in the world, but it can stand neither drought nor cold. It sleeps alone in a burrow dug six feet under the ground. Following a four-month gestation period, the female always bears four pups in the burrow. The pups' carapace appears a few days after birth.

THE RABBIT

The rabbit lives in a den. At the slightest hint of danger, it strikes the ground with its hind legs to warn the other rabbits, and then it flees to shelter. Inside the den, following a one-month gestation period, little furless rabbits are born on a hair bed prepared by their mother. The female suckles the young for 20 days. She is very prolific, having four or five gestations every year, bearing two to seven young at a time; a female may have more than 200 offspring in six years. At dawn and at dusk, all these rabbits gather on the meadows to eat grass and lick the dew on the plants. The rabbit has an excellent sense of smell, and it sniffs all the time. (This is what it does when moving its little nose.) There are many wild rabbits; man has tamed some species for meat and fur, to make them pets, and even to use them as laboratory animals, starting in the 19th century.

In Woods and Forests

THE RED DEER

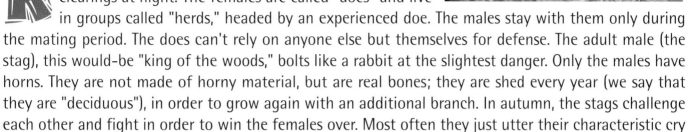

ed deer hide in woods in the daytime and feed in clearings at night. The females are called "does" and live in groups called "herds," headed by an experienced doe. The males stay with them only during the mating period. The does can't rely on anyone else but themselves for defense. The adult male (the stag), this would-be "king of the woods," bolts like a rabbit at the slightest danger. Only the males have horns. They are not made of horny material, but are real bones; they are shed every year (we say that they are "deciduous"), in order to grow again with an additional branch. In autumn, the stags challenge each other and fight in order to win the females over. Most often they just utter their characteristic cry

(called the "bell"), which echoes in the forest, or they push one another with their strong horns. In spring, the doe gives birth to one fawn, which will stay with her for two years. The brownish, white-spotted fur of the fawn will change little by little and will become as uniform as its parents' fur.

THE GREEN STINK BUG

I f you find one of these insects on you while you are walking in the forest, take care not to smash it. Scientifically known as Nezara viridula, this insect with a scale-like body is harmless. The only thing it can do if smashed or threatened is release an acrid, unpleasant smell. The stink bug in the picture is the forest stink bug; it is a phytophagous insect, which means that it feeds on plants, more precisely on the sap of plants. You should know that certain insects feed on other insects and that the stink bug, for instance, sucks the blood of mammals; this way it can transmit diseases just like the mosquito. To suck the liquids (either sap or blood), the bug pierces the stalk of the plant or the skin of the animal with its "rostrum" (a beak shaped structure). Because of this needle and its flat body, this insect looks like a real thumbtack.

THE COATI

The coati is often called "the little bear." The coati has short legs and, except for its strong claws, it has nothing else in common with the bear. The coati more closely resembles the raccoon, but it has a thinner body, a longer, bushy tail and a longer, supple snout. It uses the snout to rummage around on the ground in search of food. The coati is omnivorous; therefore it feeds on insects, small animals, and bird eggs, but it prefers fruit. It lives in small groups and is a tree-dwelling animal. It is a skillful climber indeed, and just like squirrels, it can descend trees with its head forward.

THE BARN OWL

ust like its relative, the long-eared owl, the barn owl is a nocturnal bird of pray. (It is active and hunts at night.) It has a less-severe look than its relative, probably because of its roundness: its head is round, its eyes are round, even its wings are rounded. The barn owl nests in hollow trees or in cavities; it can even dwell in your attic if it finds an entrance in the roof. Its feathers are supple, enabling it to fly without making noise. It feeds on small rodents and sometimes on little birds, frogs, snails, and, especially, on insects. It always eats on the ground. Once its food is digested, the barn owl regurgitates the nonedible parts in the shape of small balls.

THE LONG-EARED OWL

This quiet bird with big, yellow staring eyes looks as if it is coming from a horror tale. The crests on both sides of its head are actually feather tufts imitating the ears of a cat. The owl is a bird of prey that hunts at night, thanks to its very keen eyesight and hearing. Its noiseless flight allows the owl to charge its prey, which is helpless and has no chance to fly. As in most birds of prey, the female is larger than the male and is the one who rears and defends the young. She does not build a nest of her own, but lays and incubates her eggs in an old, deserted nest, in a hollow tree or on the ground, or sometimes in a burrow abandoned by a prairie dog. Did you know that the white owl (or the snowy owl) is the emblem of Quebec? It symbolizes courage and effort.

THE THRUSH

The thrush is found the whole world over; it lives especially in the forest, but it is comfortable living in parks and hedged farmlands, too. (These are fields or orchards delimited by hedges or trees). Its piping song is elaborate, and the thrush is classified as one of the best singing birds. Generally, the thrush has a spotted chest and abdomen (just like the female shown in this picture standing close to her chicks). This bird finds food on the ground: it feeds on worms, snails, insects, and fruit. Its nest resembles that of the blackbird; it is generally built among the branches of a tree, at a height of 10 to 33 feet from the ground. The female builds the nest of stems, roots, and twigs in March or April. Then, the female lays four or five blue-green eggs with reddish spots that she incubates alone for two weeks. Two weeks after hatching, the nestlings are taken care of by the male, while the female prepares for the second laying.

THE POLECAT

The polecat has brownish fur and shelters in hollow trees or in old burrows deserted by other animals. The polecat is more active at night and eats every kind of food it finds: insects, mice, small birds, leaves, and roots. You'd better stay away from this animal! A polecat can spray its enemies with a very ugly-smelling liquid produced at the base of its tail. But did you know that it can direct this jet with amazing precision to a distance of 10 feet and that its smell is so strong that the wind can carry it one kilometer away? To spray this liquid, the polecat stands in a "U" position, so that both head and tail are directed toward the enemy. This liquid represents its fiercest weapon, and because of it predators tend to avoid the polecat.

THE HEDGEHOG

Although covered entirely by spines, the hedgehog is gentle and timid. When threatened, the hedgehog straightens up its short spines and rolls itself into a ball. However, the fox has found a trick to "unfold" it and eat it: the fox covers it with urine to make it unfold and then catches it by the abdomen, the only spineless part of its body. The hedgehog lives in a family and hunts at night: it feeds on the so-called harmful animals (mice, field mice, snails, worms, insects), thus helping gardeners a lot. It also feeds on vipers, whose bites it is not afraid of. During its night tours, the hedgehog frequently crosses roads, moving slowly and stopping when blinded by car lights. In winter it sleeps in a den dug under a layer of fallen leaves.

THE MOOSE

The moose is the largest mammal of the Cerdivae family. Its big ears, thick nostrils, and especially its fleshy upper lip give it quite a strange look and a rather foolish expression. The male is larger than the female, and a big, hairy pendant of skin hangs below his throat. The adult male has flat, palette-like horns that may be 40 inches long. The moose lives in damp forests in the northern regions of Europe, America, and Asia. It feeds on leaves, ferns, and moss. It also likes water lilies and reeds found in the middle of swamps. The moose finds it very easy to eat these plants, because it can eat them while standing in an upright position. Since its front legs are longer than its hind legs, it must kneel to drink water and to graze if the plants are on the ground.

THE WOLF

he wolf lives in very well organized packs. The largest and strongest male is the leader of the pack; the other males come next in this hierarchy, then the leader's female, the other females, and eventually the pups, in order of strength. The wolf is myopic; it cannot distinguish its own companions from a distance of 20 inches. However, its eyesight is much better at night than in the daytime, and this is when it goes out hunting. Instead, it has a very keen sense of hearing. But among all its senses, the most useful is the sense of smell; the wolf is capable of smelling its prey from one mile away. Immediately after the prey is located, the pack chases it until it is exhausted. The wolf runs very quickly and noiselessly, because it runs on tiptoes.

THE LYNX

This extraordinary climber with penetrating eyesight lives as perfectly well in the woods as on the plains and in the mountains. Lynx resemble large cats; they have a characteristic tuft of thick, black hair on the tip of each ear, and their teeth are obviously stronger than those of most cats. Lynx are patient and careful animals that hunt at night. Their prey consists of field mice, deer, chamois, skunks, and sometimes sheep. The lynx is a solitary animal and couples only during the mating season; the pair stay together for a very short period of time, because the female nurtures the two to six cubs alone. The young lynx leave their mother at 10 months and can live for 15 to 17 years. The enemies of the lynx are pumas, wolves, and man, who hunts it for its soft, silky fur.

THE FOX

The fox is an elegant, careful, and highly cunning animal. To escape from its enemies, it exhausts them by making them follow it through areas full of obstacles; then, it gets rid of its adversary by hiding in a swamp or jumping in a little hiding place. The red fox is a nocturnal animal: it sleeps in the daytime and hunts at night. It lives alone until the mating season. Shortly before the birth of six to 12 kits, the parents build a burrow or use the old burrow of a badger. The mother suckles the kits until they are one month old, after which they are fed on digested and regurgitated food from their parents. At four months of age, the kits leave their parents and separate to live alone. Foxes can live eight to 12 years, unless they catch rabies, are hunted down by stronger animals (such as the wolf, the wild dog, or the lynx), or are killed by man for their splendid fur.

THE WILD BOAR

This hunchback of the woods is the wild relative of the pig, but unlike the pig, the wild boar does not spend its time wallowing in mud. It bathes in mud sometimes, too, but only to cool off or to clean its skin of parasites. As in pigs, the muzzle of the wild boar ends in a snout with which it routs along the ground searching for food: roots, tubers, small animals, and carrion. The female (known as a "sow") can bear two to six piglets. They have striped skin until the age of six. A piglet always sucks milk from the same teat. Be careful; if you meet a wild boar in the woods, you should know that when grinding its teeth (we say that it "breaks the nuts"), it means that it is ready to attack... and beware... a large male can have 330 pounds of muscle!

THE FALLOW DEER

This graceful animal has reddish-yellow fur, often spotted with white. The fallow deer is a cud-chewing animal and is related to the stag, the roebuck, the reindeer, and the elk. The male has a pair of horns that are shed and regrown every year, and that are covered by a special skin called "velvet." The ends of its horns are flat. Shortly before the horns are shed, the fallow deer rubs them on trees or bushes to remove the velvet in strips. The males and the females live separately, and they join together during the mating season. The female bears one fawn (kid) generally, but sometimes two or three. The skin of the fallow deer is highly valued by man, because it turns into very soft, good-quality leather after processing.

THE WOODPECKER

The woodpecker is black and white. It has a nice feather crest and "mustache," which are red in the male and black in the female. The woodpecker is a faithful partner. Indeed, a pair never separates; once formed, a pair stays together forever. The pair nests in a large tree hollow. The male and the female take turns incubating the three or four eggs. Sometimes the woodpecker can hardly be seen; however, it can be heard from great distances because of the characteristic "tap-tap" sound that echoes in the forest when it taps the wood of trees with its very hard beak. This way it breaks into tunnels dug by larvae or ants under the bark of trees. Then, the woodpecker stabs the insects with its long sharp tongue and eats them.

THE BADGER

This black-and-white-headed animal lives in a family made up of five to 12 animals. The badger is a nocturnal and burrowing animal; it feeds on little animals (earthworms, moles), fruit, or honey. It has odorous glands with which it marks its territory and imbrues its fellow creatures with its characteristic smell during "back-to-back" recognizing sessions. Asian badgers are able to secrete a repulsive smell. In spite of this, the badger is a very clean animal: its burrow and surroundings are always remarkably clean. Its complicated burrow consists of several galleries and exits that allow both ventilation and escape in case of danger. Some burrows have resulted from the labor of several generations of badgers and are real mazes that can reach depths of 13 feet and widths of 1.2 miles.

With its thick tail, like a bunch of feathers, this small, black-and-white, cat-sized animal reminds us of a stuffed animal or a cartoon character. But don't be fooled; the skunk is related to the polecat and just like the latter, it has two little glands at the base of its tail that produce a repulsive-smelling "musk." If threatened, the skunk can spray this ugly-smelling, oily, irritating liquid toward the enemy. Farewell to happiness! Fortunately, the skunk uses this defense as its last resort. Indeed, when disturbed, the skunk growls or whistles first and beats the ground with its front paws. The skunk is a nocturnal animal: it feeds at night by routing through the ground with its long claws, searching for insects, larvae, plants, or small animals. By the end of fall, the skunk has stored serious fat reserves under its skin; then, the female looks for a deep burrow, where she will spend the winter accompanied by the four to six kits, which were born in spring.

THE SKUNK

THE SQUIRREL

A skillful climber with charming eyes and thick tail, this little rodent is the acrobat of the forest. It has a restless and quarrelsome character. It is both fearful and solitary, but it fiercely defends its territory. It feeds on young shoots, bark, hazelnuts, and acorns, and sometimes it steals eggs from the nests of birds. In summer it builds food reserves; it can stock up almost 220 pounds of food that it buries almost everywhere; the squirrel will remember these places and will dig there in winter when food is scarce. The squirrel builds its nest in a tree hollow, high enough to shelter from predators (the lynx, the fox, and the owl). Its spherical nest is made of twigs and leaves and is covered with moss and feathers. The female bears four to six young in this nest.

In the Countryside

THE CHICKEN

The hen is the most widespread farm animal in the world. Its meat and the eggs it lays almost daily represent the main source of protein for many people. Unlike the hen, the rooster is very colorful, and his scarlet crest is more prominent than the crest of the hen. You surely must have been awakened by its famous "cock-a-doodle-do", which is much more impressive than the clucking and cackling of a hen. The rooster is aggressive and does not easily tolerate the presence of another rooster in his henhouse; this inevitably leads to quarrels that end with powerful strikes of beak and spurs (the spurs are long, horny outgrowths on the legs, which you can see on the cock in the photo). When hens are ready to breed (we say that they become "broody"), they incubate a dozen of their eggs for 21 days; the chickens hatch by breaking the eggshell with their beaks. The hens and the rooster enjoy eating earthworms, corn or wheat grains, grass, and tiny insects. While pecking, they also swallow gravel; once in the gizzard, the gravel helps to grind the grains.

THE COW

The cow moos or lows. It is a real milk factory. Did you know that some milk cows can produce up to four big buckets of milk every day? Milk is good to drink, but it also can be used to prepare yogurt, butter, fresh cream, ice cream, and countless sorts of cheese. Milk is also useful because it serves as food for the little calf that is born after nine months of gestation (just like in humans). The cow is a cud-chewing animal, and just like all ruminants, it does not chew the grass it grazes, but swallows it. The grass accumulates in its stomach and then it goes up from the stomach to the mouth to be chewed, or ruminated, before being swallowed for good. The grass is digested in the paunch of the cow (we call one compartment of the cow's stomach the "paunch" or the large "rumen") from the action of certain bacteria specialized in destroying vegetal cells.

THE DUCK

All ducks, either wild (like the gorgeous wild duck pictured in the inset photo) or domestic (like the white duck to the right), have webbed feet and waddle in a funny way along the shores of ponds or pools. The female lays about 10 eggs in a nest on the ground, which she incubates for 28 days. When they come out of the eggs, the ducklings are covered with yellow or brown down. If the mother is not present when the ducklings hatch, they will adopt any other animal and will follow it just like their own mother (if you are around, they will adopt and follow you everywhere). But generally the mother is present all the time: she is a good mother and stops often to gather her troop. The duck always tries to repel intruders, either man or animal, that threaten the safety of the ducklings.

THE TURKEY

The turkey is a massive bird whose weight can reach 44 pounds. It can fly only short distances. It feeds mostly on grains, acorns, and fruit, but it also enjoys locusts and spiders. The males fight among themselves during the mating season. When the wattle of a male (the double fold at the base of the bill and on the neck, which you can see in the two pictures above and below) reddens intensely and the top of its head turns blue, it means that the turkey is ready to fight. The one that wins this contest displays his plumes like a peacock, runs, fidgets, and clucks until its chosen mate surrenders, or rather its chosen mates, because the male has a harem of several hens. The turkey hen lays between eight and 15 eggs in a nest on the ground; it incubates the eggs for 28 days. The turkey and the hen are highly valued for their feathers (we make mattresses and decorations with them) and especially for their tasty meat, which is often cooked for holidays.

THE PHEASANT

The male pheasant is a very beautiful bird with colorful plumage. To court a female, it displays its long splendid tail. The female is smaller and drab. Its plumage blends in perfectly with the tall plants and bushes where it lives and nests. In spring it lays and incubates about 15 eggs. Pheasants are timid. If they are taken by surprise, they hide in tall grass and

remain there motionless and invisible. Only when you are about to step on it will the pheasant fly up suddenly, uttering loud cries of fear. The pheasant feeds on grains and young plants and causes great damage to the cultivation of corn, wheat, cabbages, and strawberries. However, this damage is compensated for by the fact that it also feeds on small harmful animals like slugs, snails, and insect larvae.

THE SHEEP

Shepherds raise sheep for wool and meat. The female gives very sweet milk that is good to drink and to produce cheese. The male-called the ram-has curled horns; except during the reproduction period, the ram lives separately from other sheep and lambs. In winter, they remain in the sheep barn, where they feed on hay and grain; in spring, they graze in pastures. On the hills near mountains, you can still meet shepherds leading huge flocks of thousands of sheep to their alpine pastures. This practice is called "transhumance." The shepherds shear the sheep before summer so that they will suffer less from the heat, like the sheep you can see at the bottom of the next page (one sheep provides between 2.2 and 13 pounds of wool). There also are wild sheep, like the mouflon, that live in high mountains.

THE DONKEY

The donkey has quite a sad look, but it does not like solitude. Everybody knows its cry: we say that it brays. The donkey is distinguished from its relative, the horse, by its bray, small size, long ears, and gray color. After less than a one-year gestation period, the female gives birth to one donkey foal. This adorable young foal will have soft skin until the age of two. Did you know that if you cross a donkey with a mare, their offspring is called a "mule"? Crossing a stallion with a she-ass is much rarer, and the offspring is called a "hinny." But neither mule nor hinny can bear young (they are sterile). Since Greek antiquity up to the 21st century, people have used the donkey to work the land, to transport heavy loads, or to pull carts. Today it is used to carry tourists and as a pet.

THE GOAT

When the goat utters a "baa," we say that it baas or bleats. It is very cute, with its small goatee beard, but pay attention when you enter a pen where there is a goat, because it can eat anything. And if you let it, the goat will try to chew your purse or clothes. The male has longer horns than the female and when angry, he charges with his horns straight forward. The female bears one or two young (which are called "kids") that she will suckle. Goat milk is quite fatty and has a characteristic taste and smell. It is used just like cow's milk to produce delicious cheese. Some goats have very smooth, highly valued wool (cashmere); others are wild and live in craggy mountains.

THE HORSE

The horse has accompanied man for almost 5,000 years in his everyday life (for transportation and agriculture), but also in dramatic situations (on battlefields!). There are countless species of horses, adapted to various activities: the racehorse, the draft horse, and so forth. And don't forget the pony, which is the smaller brother of the horse. Do you know why people shoe horses? Because the shoes protect their hoofs; they lessen wear and keep infections away. The female is called a "mare" and the male a "stallion." After 11 months of gestation, the mare bears a baby called a "colt" or a "filly" (depending on whether it is male or female). The mare suckles the young for almost six months. The horse sleeps eight hours a day, but not for long periods; most of the time it sleeps in an upright position.

THE PONY

The pony is a small horse that stands just under 15 hands high (traditionally a horse's size is measured at the withers - the elevated part of the spine between the neck and the back - and the measurement is made in hands). All those exceeding this limit are called "horses." The pony neighs and is herbivorous just like the horse; yet, its legs are shorter, its thick tail almost reaches the ground, and it has a bushy, tousled mane. The Shetland pony is one of the smallest ponies; it stands 10 hands tall. Despite its small size, the pony can pull a load two times heavier than its own weight (that is a little more than 2200 pounds). In fact, it was used in the past in mines to pull ore or coal wagons. The female (the mare) carries her young for 11 months. The foal is able to stand very soon after birth; at the age of one month, it starts grazing grass just like its mother.

THE PIG

The pig is a round giant with a little curly tail. It eats a lot and almost every kind of food. (It therefore is omnivorous.) Its grunting and squealing produce a lot of noise. A large boar (or uncastrated hog) can weigh up to 880 pounds. The sow bears about 10 piglets that continuously suck milk for 21 days; therefore, the 12 teats of the sow are often not enough to feed this little population. Almost everything is used from a pig: the meat, the skin for leather products, and even some hair for brush manufacturing. It is also very useful in medicine; its skin is used as transplanted tissue in serious burn cases. The pig has a remarkable sense of smell; we use pigs to find highly prized truffles (mushrooms) under the ground.

THE LOCUST

The locust resembles the grasshopper; it is also called the "short-horned grasshopper," because its antennae are always shorter than its body. Its hind legs are very strong and help the locust jump very far. It has two pairs of wings, which enable only short-distance flight for all except for the migratory locust. The locust produces a sharp sound by rubbing together its hind legs, which are provided with prickles, or by rubbing its forewings against other parts of its body. This is why we say that it stridulates. The locust feeds on grass and leaves. At the end of the fall, the female deposits her eggs in a special tube on her abdomen (called an "ovipositor"). The next spring the larvae will come out of the eggs. Each larva sloughs four to six times before transforming into a little locust that will become an adult that very year or the next. Its color helps it hide efficiently to escape its numerous predators (birds, little carnivores, insects).

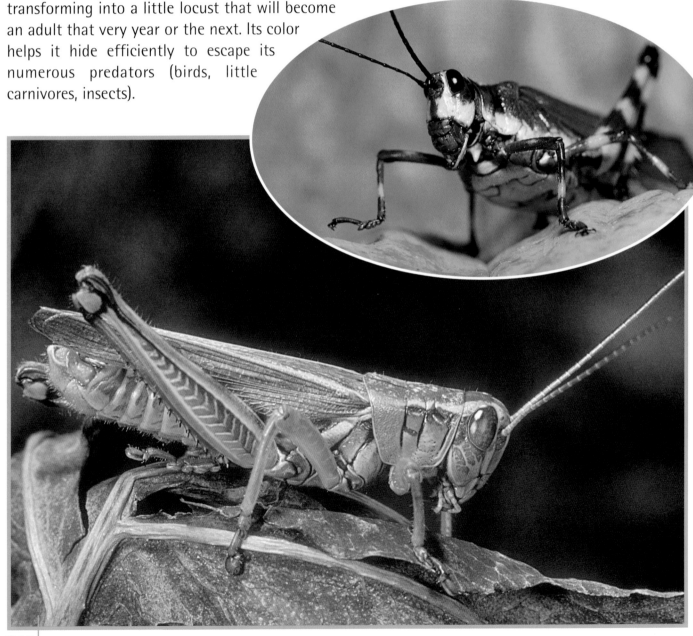

THE GRASSHOPPER

There are around 15,000 different species of grasshoppers; the majority live in tropical regions. The grasshopper is an insect you can meet everywhere in meadows. While walking through the grass, you probably have seen grasshoppers jumping at every step you made. The grasshopper is distinguished from the locust by the length of its antennae, which are longer than its body. The grasshopper is primarily carnivorous and feeds mostly on little insects. You can also distinguish it easily by the chirping or stridulating noises it produces by rubbing its wings together. The female has an ovipositor at the extremity of her abdomen; this is a kind of a flat tube enabling her to lay her eggs on the ground or in a stem. The eggs hatch the next year. A wingless larva comes out of every egg. Before it becomes an adult grasshopper, the larva will undergo repeated sloughing for a period of three months.

THE ROOK

ust like all ravens, the rook's plumage is completely black. It is distinguished by the black feathers covering the upper part of its legs and by the whitish patch around the base of its bill. It is a very gregarious bird that forms noisy flocks. It lives in fields, where it scratches the ground in search of shoots, worms, insects, and little rodents; sometimes the rook also feeds on carrion. The rook nests in colonies that are known as "rookeries." The nest is made of twigs in a tree where there already are several other nests. The female lays three to five eggs that she incubates alone for 15 days. After that, the male helps her care for the young. When leaving the nest, at the age of one month, the chick does not have the characteristic white patch around the base of its bill and is sometimes mistaken for a carrion crow. However, if you look at them carefully when they are on the ground, you will notice that the carrion crow jumps while the rook walks!

THE RAT

The rat looks like a big mouse. The size of its body varies from eight to 10 inches, and its tail can be almost eight inches long. The rat is found especially in places close to human dwellings: in cellars, barns, garages, stables, and piles of timber. But it also lives in meadows and in forests on all the continents. It is an omnivorous rodent; it feeds on small birds, eggs, seeds, even garbage, and it can damage food supplies. It spreads its excrement everywhere, and this is why it can be a disease carrier: we must beware of it! It can be aggressive if threatened; it doesn't hesitate to attack its enemies with its teeth and claws. The female can have 12 gestations per year and can bear about nine young every gestation, which makes her as fecund as the female mouse.

THE PEACOCK

The tail of the male is made of gorgeous, huge blue and green feathers. When the male wants to be noticed by a female, it majestically displays its fan-like tail, forming its splendid, famous "wheel" like in this picture. Nobles valued its meat highly in the Middle Ages. Today it is domesticated especially for its beauty as an ornamental bird. The peacock, which can be seen in public parks and gardens, originated in the forests of Asia. This bird normally feeds on seeds and insects, but if hungry, it can kill a snake with beak strikes in order to eat it. During the mating period, the male is surrounded by four or five females; they can easily be distinguished by their drab plumage. Generally, the peacock builds its nest on the ground, and the female lays eight to 20 eggs that she incubates for 32 days. They can live from 10 to 15 years. Their call is a loud scream: "Kee-ow," "Kee-ow," like the meow of a cat.

THE FLYCATCHER

This is a familiar bird that hardly draws attention. The colors and the look of the flycatcher are very common, and it is very discreet. It is approximately the size of a sparrow. You can meet it in the countryside, in woods, in parks, or even sitting on a power cable or on a pillar. Its name suits it perfectly; it is a fearsome insect hunter. As soon as a fly, a butterfly, a wasp, or a dragonfly comes close without noticing it, the flycatcher charges in for the kill. The flycatcher builds its nest, using stems, lichen, and moss in a wall, in an ivy bush, and sometimes even in an old nest if the opening is wide enough. The female lays four or five greenish-white eggs with violet and brown spots. The parents take turns incubating the eggs for two weeks. After hatching, the chicks are cared for by both parents for another 15 days.

THE SWALLOW

The swallow is a migrating bird; its return announces that spring is coming. Although it does not weigh more than 0.7 ounces, this bird can cover a distance of 6,200 miles during migration. The species shown in these two pictures is called the "house martin." It nests in colonies, in the country, in cities and villages, and often on rocky walls. Both parents work on building the nest, which is made of little rocks and straw cemented with clay soil mingled with saliva. These nests are highly valued in Chinese cookery, where they are made into a famous soup. The female lays between three and five white eggs; the two parents incubate them for two weeks. The chicks leave the nest three weeks later, but the family stays together until the chicks become completely independent. The swallow is a true master of flight: it twirls in all directions, making incredible turns. It hunts insects, which constitute its main diet, while in flight.

THE GOOSE

The goose is an excellent "guard dog"! A flock of geese becomes terribly noisy when the most insignificant intruder shows up. And woe betide anyone who is attacked by geese; their beak pecks are very painful. When eating grass, the goose actually mows the lawn with its indented beak. The white goose shown in these photos has been domesticated since antiquity for its meat and eggs. Its feathers once were used as a writing implement. People also fatten geese to obtain foie gras (a sort of liver pâté) that is so much appreciated during the winter holidays. The female is called a goose, of course, but do you know the name of the male? He is called a "gander." As for the goose chicks, they are called "goslings." The chicks come out of eggs in a nest made of wood and plants and covered with down; the parents rear the goslings together. The goose can live 20 years. Wild geese are migrating birds; they fly from one place to another, depending on the season. They fly in flocks, forming the characteristic letter "V" in the sky.

In the Garden

THE BEE

The domestic bee lives in a colony called a "swarm" that nests in a shelter called a "beehive." The nest is made of perfectly hexagonal wax alveoli (like those in the picture below). The colony is very well organized; it consists of a queen, a few males (the drones), and thousands of workers, which are the sterile females (they cannot breed). The golden rule of the colony is "everyone mind its job"! The males inseminate the queen, which lays one egg per minute, and the workers produce, maintain, and clean the swarm, leave in search of food (pollen or flower nectar), and look after the queen's young. Once laid in an alveolus, each egg will transform after 21 days into a worker bee that will then leave to collect nectar. The worker bee brings the nectar to the swarm, deposits it in an alveolus, and then covers it with wax. Of course, honey is obtained from this nectar! When a worker finds a place with many flowers, it announces this to all its fellows, performing a sort of dance to show them where the place is.

THE LADYBUG

The ladybug is a nice, tiny coleopteran insect, often colored red, orange, or yellow, with black spots. You surely have played with a ladybug by keeping it on your hand until it flies away. The ladybug is a very useful insect. It protects the plants in our gardens by eating a series of insect pests (plant aphids, dust mites, cochineal insects). It is estimated that a ladybug can eat about 3,000 plant aphids during its short lifetime. It also feeds on leaves. The ladybugs mate in spring. The female releases a characteristic smell to stir the males. The female dies shortly after laying the eggs (three to 300 eggs, depending on the species). In four or five weeks, a larva will come out of the egg; then it will change into a colored nymph and finally into an adult ladybug. The latter does not live more than a year.

THE BUTTERFLY

There are at least 150,000 species of butterflies all over the world. The butterflies that you see in gardens are diurnal (like the ones in these pictures), which means that they are active in the daytime. The vivid colors of their wings are caused by the little scales that cover them. These scales produce the "powder" that remains on your hand after having touched a butterfly. The butterfly feeds on flower nectar, tree sap, or fruit juice by means of its long trunk. When the butterfly rests, the trunk is curled (as you can see in the picture at right bottom). The butterfly transports pollen while flying from one flower to another; it thus contributes to the reproduction of flowers. The female lays eggs, which hatch after one or two weeks, and the caterpillars come first. After a period of 10 days to one year, the caterpillar transforms into a butterfly. Depending on the species, the butterfly lives from a few days to six months, seldom more. Some butterflies hibernate under leaves or in a hole during winter; others can migrate far away. (Some of them can even reach Africa!)

THE SNAIL

The snail is a gastropod mollusk (literally belly-footed animal); therefore, it is a distant relative of the oyster and the octopus! Its shell serves as shelter; the snail withdraws inside at the slightest danger. Study these photos and you will see that the snail has two pairs of antennae. The pair of longer antennae (also called "horns") each end in an eye. The two smaller antennae act as both hands and nose: the snail can touch and smell with them. Snails move by gliding on a track of deposited viscous substance called "slime." The snail hides inside its shell when it is too hot, so that it will not dry out; it shuts the shell with a kind of lid produced from the same slime. Snails feed on plants that they "graze," their tiny teeth forming a sort of grater. When fall comes, the snail hides under the ground, where it sleeps all winter long. The snail is a hermaphrodite, which means that it is both male and female at the same time. After mating, each partner leaves and lays eggs on the ground; a few weeks later, tiny snails come out of them.

THE ANT

nts are gregarious insects, just like their relatives, the bees. They live in very well-organized societies that sometimes number tens of thousands of individuals. Everything starts with a mating dance; the queen takes off and is followed by a horde of winged males. Once inseminated, the queen loses her wings and starts accomplishing her main role, that of laying eggs. All the females born from these eggs are sterile and cannot lay eggs; they are either worker ants that perform most of the work of the colony or soldiers that defend the colony. The anthill spreads little by little in a giant network of galleries and rooms. The workers take care of the eggs and the larvae and nourish the colony. Some ants breed leaf aphids (from which they produce a sort of sweet liquid that they feed on), others cultivate mushrooms on a layer of leaves, which they transport in pieces (this is what the ants below are doing); others leave to find food or simply steal the food of neighboring anthills.

THE WASP

With its reputation as a "stinging" insect, the wasp is not very popular. However, it usually stings only when threatened (yet, it is true that it is very nervous). Therefore, if a wasp is flying over your plate, you best ignore it; just don't agitate it. And, especially, shut your mouth, unless you want to swallow it inadvertently. (It is very dangerous to be stung inside the throat.) There are some wasps that prefer to live alone, but most of them are sociable: they live in groups like the bees. After winter is gone, a female starts building her nest with a paste produced by chewing wood and combining it with saliva. After that, the female lays her eggs in the cells of the nest. Each egg will bear a little white larva that will transform into a wasp. A nest can shelter several dozen to more than one thousand individuals (as with the social wasps).

THE PIGEON

In antiquity, the Romans used to fatten pigeons to eat them, while knights in the Middle Ages used them as messengers. Pigeons have an extremely well-developed sense of direction. Even nowadays, the "carrier pigeon" (like the ringed pigeon in the picture below) is used in long, point-to-point flights. It is capable of flying more than 20 hours without a break. Pigeons are found almost everywhere; you can hear them cooing in cities, in public parks, in the countryside. Sometimes a pigeon can cause significant damage. In the countryside, for example, it bothers farmers a lot, because it feeds on their crops (peas, rapeseeds, sunflowers, wheat). It nests in trees, under the slopes of roofs, or even on statues. Its nest is a small platform of twigs. In spring, the female lays two little white eggs. She can have a second breeding at the beginning of summer and sometimes even a third breeding early in fall. If a pigeon can live 10 years, imagine how many chicks it can breed in its whole life!

The slug belongs to the group of mollusks. It closely resembles the snail, but generally it does not have a shell on its back. (Some slugs do have a shell, but it is not visible to the naked eye.) It especially loves damp places, and you can often see it in fields or along roads when it rains or after it rains. You can come across it on the ground, under a leaf, on tree bark, or under a stone. As it does not have a shell to protect it against drought, it continuously secretes a layer of slime that covers its body and protects it against drying out; this layer of slime gives the slug its sticky look. The trace it leaves when moving is made of the same slime. The slug feeds on greenery and can cause serious damage in flower and vegetable gardens.

THE SLUG

THE MILLIPEDE

O f course the millipede does not have 1,000 legs, but it does have a lot of them. It is found almost everywhere in the world, regardless of cold, hot, dry, or humid weather. The millipede and the scolopendrid, its relative, belong to the class Myriapoda, which is related to the insects. The length of its body ranges from two to eight inches and is formed of a succession of segments. It may have up to 80 segments, and each segment is provided with a pair of legs. A millipede can have about 320 legs. Right, that is quite a lot! The millipede also has two antennae on its head, but unlike the scolopendrid, the millipede does not have venomous fangs. During the reproduction period, you can often see two linked millipedes, just like in the photo below. If threatened, the millipede coils into a spiral and stays still. It is a vegetarian and can cause damage in gardens.

THE BLACKBIRD

The blackbird is one of the largest passerines, whose body can be 10 inches long. The common blackbird (the male has black feathers and a yellow beak like the one in the top picture; the female is brownish like the one in the bottom picture) and the American blackbird (which has a gray-brown back and an orange chest) are the best-known species. The song of the blackbird is sharp, melodious, and distinctive. This bird feeds mainly on earthworms and insects; in summer it also enjoys fruit. It is the naughty bird that eats the cherries just before we pick them. In spring, the male chooses a territory, and the female builds a nest there from twigs and grass cemented with mud. She generally builds the nest in bushes or hedges or attached to a building. The female lays three or four greenish-blue eggs and incubates them for 12 to 14 days. The young are ready to leave the nest at two weeks old. The father accompanies them for two more weeks, while the mother incubates more eggs. Unless it ends up in the belly of a cat, a snake, or bird of prey (the crow, for example), the blackbird can live 14 years.

266

THE STARLING

The starling is a common presence in gardens, where it finds its natural diet (insects, worms, and fruit) and whatever is offered by people (bread, seeds). It lives, moves, and sleeps in groups. When in flight, the starling is distinguished by its sharp, triangular wings. In spring, the male sports mating plumage that is black with metallic blue, green, and violet highlights; in fall, its plumage resembles that of the female and is spotted with white. The male sometimes is polygamous. The female often nests in a tree hollow, where she stacks bits of straw, wool, and feathers. She lays about three or four light-blue eggs that she incubates for 14 days. The chicks take off at the age of three weeks. Starlings generally migrate; there are, however, several types that winter in cooler regions. The song of the starling is a continuous chattering accompanied by whistling and creaking. It partially imitates various sorts of sounds (the cry of owls, of curlews, the ring of the phone).

THE ROBIN REDBREAST

This 5.5-inch birdie is lively and quite round in shape. It is distinguished by the red color of its breast, hence the name of "redbreast." It feeds on all kinds of insects and spiders it finds on the ground. It is found in cities as often as in forests. It mostly prefers to stay on the low branches of trees and to sing its nice melodious song. It is very comfortable in gardens, and sometimes it comes to peck at the grain people leave on their windowsills. A pair forms during the mating season. The nest is made of moss and dry leaves and is generally built in a hole in a wall or a tree, and sometimes even on the ground. The female lays five to seven eggs in a nest twice a year and incubates them for 13 days. The male feeds her all this time. When the young hatch, the parents take turns feeding them for three weeks.

THE SPARROW

The sparrow is about 5.5 inches long. It belongs to the passerine group. It feeds on grain, earthworms, and insects. It is impossible not to associate this bird with people. It is always present in places where people live. It accompanies people even at an altitude of 7,500 feet! It lives in flocks, and in summer it finds its diet in fields of grain (wheat, oats, barley). The male chooses his dwelling in a hollow in a wall or a roof beam and defends his territory fiercely. Then, he lures a female by chirping and by jumping noisily around her. In April, the female lays four to six light-colored eggs that are spotted with gray and dark brown. Both parents take turns incubating the eggs for 12 to 14 days. The young start flying at the age of two weeks. They must be wary of cats, which also live around people.

THE EARWIG

The earwig is distinguished by the "forceps" at the end of its body, which it raises whenever threatened. Its name comes from a legend that says if an earwig enters our ear, it is capable of breaking our eardrum with this famous pincer. Of course, this is not true at all, and there is no reason why you should fear this completely harmless insect. Its pincer is formed of two jaws that are curved in the male and almost straight in the female (as is shown in this picture). The earwig feeds mostly on vegetal residue, vegetables, and flower petals. It stays out of the light and generally lives hidden in vegetation or in houses. The female lays between 20 and 40 eggs in the ground; then she takes care of them all winter long. (This is an entirely exceptional behavior in insects.) The larvae come out of the eggs by the beginning of spring and are nourished and reared by their mother; they become adults by the end of summer.

THE WOODLOUSE

Don't let yourself be fooled; the woodlouse is not an insect, but is a relative of the crab and the shrimp. Therefore, it is a crustacean. It is actually the only terrestrial crustacean. It has seven pairs of legs for walking and needs much humidity. You can find it under a pot of flowers that have been abundantly watered, under a damp fallen branch, in the garden compost, or in a cellar. If you touch it, the woodlouse turns into a ball and stays still for a few seconds. The woodlouse is a very useful animal; it feeds on vegetal residue and changes decomposing dead plants into useful substances for new plants. The female carries her eggs in an incubation bag under her thorax. The larvae emerge

from eggs in this bag filled with liquid and stay there for several weeks; 24 hours after coming out of the bag, the larvae start sloughing. Woodlouses live about two or three years; they are preyed on by millipedes, lizards, toads, and some spiders.

THE SCARAB

The scarab belongs to the order Coleoptera, which numbers more than 19,000 species. The June bug, a common name for any of several beetles in the scarab family, and the rhinoceros beetle, presented in the small picture to the right, belong to the same family. The common beetle (presented in the large picture) is black and sometimes is iridescent with metallic hues. This is a dung beetle, which means that it uses the excrement of cud-chewing animals, for example, the cow. It makes pellets out of the excrement (sometimes bigger than its own body) and pushes them long distances by walking backward to its gallery. These pellets provide the diet of both the beetle and its larvae. The female lays eggs in this gallery. The newborn larvae find the food at hand here. The larva transforms into a nymph and then into an adult beetle. In ancient Egypt, the scarb was the symbol of resurrection, and, thus, little amulets in the shape of beetles were placed in sarcophagi.

THE MOLE

The mole is a small mammal very well adapted to underground life. Did you know that moles have very poor eyesight? Some of them are completely blind. We should mention that eyesight wouldn't be very useful underground. Instead, the mole has excellent senses of touch, smell, and hearing. The mole digs extensive galleries with its flat, robust, short paws, which end in strong claws. It is a real "mechanical digger." It is able to dig a six-foot tunnel in 10 minutes through soil of normal consistency! Of course, it does not dig for pleasure, but for food; its diet consists of insects, larvae, and earthworms. It is itself the prey of owls, foxes, and snakes. Man also hunts it to prevent the damage it causes in gardens and for its nice fur.

At Home

THE CAT

The cat we see in our homes every day is a feline. Well... that is perfectly true; it is the younger cousin of the lion and the tiger. This fact is evident when a cat hunts a mouse; it stalks noiselessly, just like its wild relatives. It is the most widespread feline in the world; you can meet it anywhere on the entire planet. It has accompanied humans for a long time. In ancient Egypt, the cat held an important place among the gods. For instance, the name of the Egyptian cat goddess of music, dance, and motherhood was Bastet. If you want to adopt such a "god," you should wait until it is at least eight weeks old and is totally independent. The cat is carnivorous. It purrs like all felines. It manages to produce this characteristic sound by way of several coordinated movements of the larynx and the glottis. Purring generally denotes contentment and pleasure. Sometimes it can also express fear and stress.

Don't forget to treat your cat gently; unlike dogs, the cat has no master and is affectionate only if it thinks you deserve affection. Remember to take good care of this pet! Although it is a very independent animal that can manage all alone, various diseases can endanger its health. Make sure especially that it doesn't have fleas. Flea "bites" are not only annoying, but they can also transmit parasites to your cat, especially a flat worm resembling the tapeworm. If that happens, you will have to administer anthelmintics (substances that kills parasitic worms) to your cat. If a mother cat becomes infested, you should give these substances to her kittens as well. The "roundworm" is another worm transmitted from mother to kittens through suckling. It can cause severe disease in very young cats. The cat bears four kittens at a time on the average, after a gestation period a little longer than two months. It suckles the kittens until they are eight weeks old. The cat lives about 12 years, but the longevity record is 31!

278

THE DOG

rchaeology has taught us that "man's best friend," the dog, has lived close to us for at least 10,000 years. There are about 400 dog breeds; you can see some of them in these pictures. Some breeds came from the wolf and others from the jackal. Dogs have preserved some features from their ancestors (such as their round pupils, excellent sense of smell, and long powerful fangs), but they have developed some characteristic features of their own, such as barking and floppy ears. Dogs are found everywhere in the world. They are easy to train, and they are used for many purposes. Certain breeds are more or less adapted for hunting, guarding, companionship, and for helping shepherds, the blind, first-aid workers, police officers, and so forth. Dogs are intelligent animals and often are very affectionate.

They are the most famous enemy of the domestic cat. Everybody must have seen a dog running madly to catch a cat. However, in certain cases (especially if they grow up together) the dog and the cat can get along really well and they can even become best friends. The female, called a bitch, bears between five and 10 puppies after a two-month gestation period. The sooner you start training your dog, the better. Generally, training starts when the puppy is seven to 12 weeks old, by teaching it to be clean and where to go. After that comes other types of training. Training for obedience is especially important in large breeds, such as the German shepherd, the Doberman, or the pit bull. These dogs are very strong, and their strength can turn against people later unless it is controlled. The "pedigree" breed dogs generally live shorter lives than the crossbreeds that come from the crossing of several breeds; a pedigree lives about 12 years, while a crossbreed can live 20.

THE CANARY

As the name suggests, the canary comes from the Canary Islands. It is a devoted singer, very well adapted to cage life. Actually, this is a very common bird; everybody must surely have seen a canary in a cage along with a cuttlefish bone fixed between two small bars! This bone enables the canary to sharpen its beak and to procure the calcium that is necessary for the development of its egg shells. It likes to have some privacy during the reproduction period; therefore, you should try to leave the pair alone throughout this period. The male canary can be clearly distinguished from the female by his different song; the male sings much more melodiously than the female does. The female lays two to six pale-blue eggs with little brown spots, but she lays only one egg a day. During the laying period, breeders replace eggs with false eggs; at the end of the laying period, they put all the eggs back to be incubated, so that the chicks will hatch at the same time. The eggs are incubated for 13 or 14 days, and the chicks are able to feed alone starting at the age of one month.

THE PARAKEET

his representative of the parrot family is the easiest bird to raise. This species has been tamed since the beginning of the 19th century. There are parakeets of various sizes and colors (white, yellow, green, blue, and red). The parakeet is a very sociable animal; this is why several parakeets are placed together in the same cage or aviary. You can easily domesticate a parakeet if you start

training it from birth. In the adult male parakeet, the upper part of the beak (called a "wax") becomes blue or crimson. In the female, this part is brown and gets darker in spring when it is ready to mate. Their cage should contain nests during this period, so that the parakeets will have everything they need. Then you can hope to see the female laying four to six eggs, one or two eggs a day. After 18 days of incubation, the young parakeets emerge from the eggs and are fed by the parents for one month.

THE GUINEA PIG

The guinea pig is larger and more fearful than its relative, the hamster. You have surely met this animal at one of your friends' places. It is not very difficult to raise one at your place either. It is enough if you have a cage, litter of straw, sawdust or wood shavings, and one or two accessories, such as a food tray and a nourishing bottle containing water and placed vertically. This is all a guinea pig needs, because it is not very active and it is a clumsy climber and jumper. It is herbivorous (it feeds only on plants and fruit). The guinea pig is both diurnal and nocturnal; in fact, it usually takes short breaks during both day and night. If you have a guinea pig pair, you should know that the female can have four or five gestations per year, bearing two or three pups each time.

THE HAMSTER

The hamster is a very active nocturnal animal that can live two or three years. It is not difficult to feed, because it is omnivorous. Moreover, if you want to have one of your own, all you need is a cage to shelter it in, a vertical nourishing bottle, a wheel for it to play on, some sawdust for litter, and a little bit of cotton or fabric so that it can make a soft nest. The hamster can breed from the age of two months; the mating season starts in May and lasts until September. The gestation period is very short, about 15 or 16 days, and it can bear five to nine young. It has no fewer than eight gestations per year. The female must be isolated during the gestation and bearing period, because cannibalism is rather frequent in hamsters, and the male tends to eat its offspring. The hamster lives solitarily in the wild. It mustn't be disturbed in winter because it hibernates.

THE GOLDFISH

The goldfish is a freshwater fish of the Cyprinidae family, originating in China. There are many varieties of goldfish in terms of color (yellow, orange, red, spotted) and shape. Some of them have soft, voile-like fins, others have a sort of red hood on the head (like the fish in the picture below) or they have what appear to be humps all over their body. The goldfish is omnivorous; this means that its diet consists of both greenery and tiny animals (worms, insects, little fish). This is why you should put the spawn in a separate bowl, because the adults can eat the young fish coming out of the eggs. Goldfish are social and can associate easily with other fish or with aquatic snails to cheer up an aquarium.

THE MOUSE

ice are the most familiar and the most widespread rodents in the entire world. They are crepuscular and nocturnal; therefore, they start looking for food after night has fallen. Mice are omnivorous. They are very clean and wash themselves carefully. They live in family groups dominated by males. Males occasionally fight viciously to maintain their hierarchical position within the group. Mice live about two and a half years, and they are very prolific animals. Indeed, their sexual activity is performed all year long. The female has four to eight gestations a year, bearing seven to 10 little mice each time. The young attain sexual maturity by two months of age. In your opinion, how many mice can come from a single pair in a year? A huge number. Anyway, enough to infest an entire building. Fortunately for us, mice are preyed on by owls, cats, and dogs, which limits their proliferation.

THE FLY

The fly is a dipterous insect, which means it has only one pair of wings. Its wings flap very quickly, making its flight very precise. Its head has two, big, facetted eyes with two short antennae between them. Its mouth is formed of an extensible "trunk" with which the fly licks or sucks the sweet liquids that it feeds on. The fly is a very fecund insect and lays its eggs almost everywhere... actually, it lays them on everything that can nourish its larvae so that they can grow and turn into flies later. Therefore, the fly lays eggs on food, excrement, carrion, fruit, and garbage. As a result, it can cause serious health problems to people and animals. Indeed, as it flies around everything, it can transmit the germs (and thus the diseases) that are found in excrement, for example. In some countries, flies carry severe diseases like cholera and typhoid fever.

THE MOSQUITO

In this dipterous insect, only the female bites to suck blood. The male feeds on sap and nectar. In fact, the female needs the blood of mammals to be able to produce her eggs. The female lays the eggs on the surface of a lake, a swamp, or a mere puddle. An aquatic larva comes out of each egg. You surely have noticed them; they look like little red worms that come to the surface to breathe. The characteristic noise produced by the mosquito when flapping its wings very fast enables individuals from the same species to recognize each other in order to mate. In addition to the terrible need to scratch our skin where a mosquito has bitten us, a mosquito bite can also transmit diseases. Indeed, some species can carry very dangerous germs and are responsible for transmitting diseases like malaria or yellow fever to people.

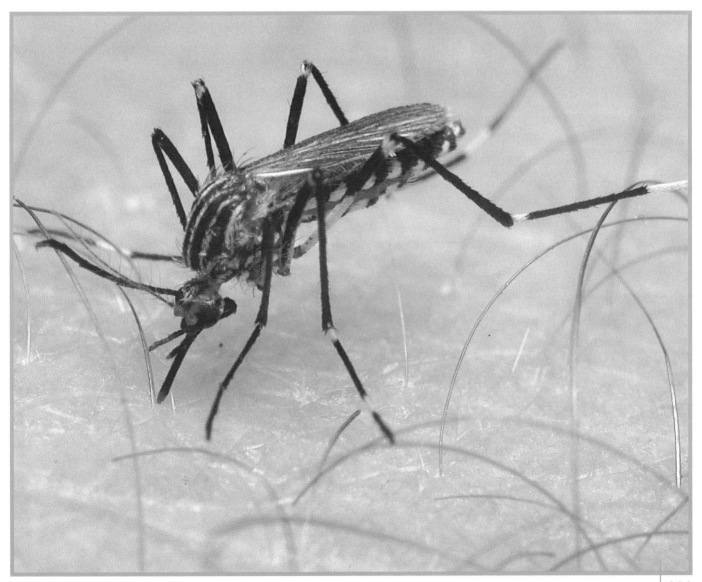

The Animal Kingdom

THE INVERTEBRATES

The word "invertebrate" denotes all animals without a backbone or spinal column. Therefore, the invertebrate class consists of such different animals as the jellyfish, the starfish, the earthworm, the octopus, the fly, or the spider. The invertebrates are the most numerous and the most diversified animals on the planet; we meet them in all natural environments (on and under the ground, in the air, in the seas, and in fresh waters). They represent not less than 95 percent of the total of over two million species of animals known on Earth. Indeed, scientists discover new species every day, and the majority are invertebrates! In establishing a classification, animals with the same characteristics are grouped by such considerations as shape (the presence or absence of a shell, the number and the shape of the legs, the symmetry of the animal, and so on), the way their organs are disposed, and even their behavioral characteristics. The invertebrate world contains 31 different, large groups, each called a "phylum." (Some of these phyla contain only a few, relatively rare animals.) The invertebrates presented in this encyclopedia belong to four of these phyla that are among the most important: the Cnidarians, the Arthropods, the Mollusks, and the Echinoderms. The Arthropod group is the largest; it consists of almost one million species, which is 50 percent of all species in the entire animal kingdom. To learn more about these animals, you can read about the main characteristics of the invertebrate groups as presented in this encyclopedia. This will enable you to acquire a basis for classifying on your own the animals you will see in books or you will meet while walking in forests or along seashores.

The **CNIDARIANS** consist of sea anemones, gorgonians, corals, and jellyfish. They are mainly sea animals, but some of them can be found in fresh water too. All these animals have in common radial symmetry of the body (the body is disposed in rays around a central axis) and the presence of tentacles, the number of which depends on the group. The phylum Cnidaria consists of three classes: the Hydrozoa (small-sized jellyfish), the Scyphozoa (large-sized jellyfish), and the Anthozoa (corals and sea anemones).

The **ARTHROPODS** represent the most important invertebrate phylum in terms of number of species. This phylum is subdivided into several classes, some of which are famous all over the world:

THE ARACHNIDS include spiders and scorpions; they do not have antennae, but they have four pairs of legs and breathe with tracheas (these are small tubes that lead air inside their body).

THE MYRIAPODS include millipedes and scolopendrids; they have a pair of antennae, a great number of pairs of legs, and they breathe with tracheas; The scolopendrids also have a pair of venomous fangs just like some spiders.

THE CRUSTACEANS have two pairs of antennae, a calcareous carapace, and five pairs of legs. They breathe with gills. They live mainly in seas (the crab, the shrimp, the lobster, the hermit crab) but some also can be found in fresh water (the crawfish).

THE INSECTS represent the most diversified group of arthropods: this class contains no fewer than one million separate species! You can identify them by the following criteria: they have segmented bodies with three main parts (head, thorax, and abdomen), they have one pair of antennae and three pairs of legs, and they generally are winged. The dragonfly, the fly, the mosquito, the ant, the bee, the ladybug, and the beetle are some of the creatures belonging to this class.

The **MOLLUSKS** are soft-bodied animals. The three main classes of mollusks are: the Bivalve mollusks, the Gastropods, and the Cephalopods.

THE BIVALVE mollusks are fresh- and salt-water animals. Their shell is made of two parts, called valves. The mussel, the oyster, and the scallop are a few examples of bivalve mollusks; their head is not separated from the rest of the body.

THE GASTROPODS are terrestrial or aquatic mollusks (they live in fresh or salt water); they have a single shell, generally spiral shaped (like a snail's) or have no shell at all (like a slug), and the head is well separated from the rest of the body. The gastropods have a lateral leg that enables them to crawl.

THE CEPHALOPODS include the octopus, cuttlefish, squid, and nautilus. They are marine animals whose shell is placed inside a wrinkle of the skin and whose voluminous head ends in tentacles.

The **ECHINODERMS** are exclusively marine animals, found in all the seas of the world and at all depths. Their body is protected by calcareous skeletal pieces, more or less developed according to the group: they can be made up of spicules (supporting skeletal structures) sprinkled in the skin or of calcareous plates that form a continuous skeleton. The echinoderms are the only invertebrates with a skeleton under the skin, like vertebrates. (Other invertebrates have external skeletons). Their spines are outgrowths of the skeleton. An identical five-part segmented body is the hallmark of this group. (We say that they present "five-part radial symmetry.") The Encyclopedia presents two of the four classes of echinoderms: the class known under the scientific name of Asteroidea (the starfish) and that of the Echinoids (sea urchins).

THE VERTEBRATES

The vertebrates emerged on Earth more than 450 million years ago. Fish were the first representatives. Afterwards, fish evolved into batrachians and the other vertebrates and little by little colonized the terrestrial environment. Some of the first vertebrates had disappeared from the surface of the Earth long before man appeared. These animals are known only from their fossils. Among these are the dinosaurs, the ancestors of present-day reptiles. In their entirety, the vertebrates represent a superclass within the phylum Chordata. These animals are characterized by the significant evolution of their central nervous system, namely of the brain and the spinal cord. The brain is protected by a skull, and the spinal cord is protected by a spinal column. The vertebrates have a heart and a closed circulatory system made up of arteries and veins. The arteries are blood vessels that convey oxygenated blood from the heart to various organs of the body, and the veins are blood vessels that carry oxygen-depleted blood from the organs to the heart. The sex of the individuals is always determined, they are either male or female. Therefore, there are no hermaphrodites in this class, and their reproduction is always sexual. The vertebrates consist of about 45,000 different species divided in five main groups:

- **fish**
- **batrachians**
- **reptiles**
- **birds**
- **mammals**

The Fish

Fish are the most ancient vertebrates; they emerged on the Earth 450 million years ago. Presently there are about 22,500 species of fish. They are cold-blooded animals: the temperature of their body is equal to that of the water they live in. All fish, either young or adult, live in water (sea or fresh water) and have gills for breathing. The gills are made up of numerous lamellae disposed on each side of the head in cavities called the "branchial chambers." In bony fishes (like the tuna or the trout), the branchial chambers are protected by an "operculum" (also called an "ear" sometimes) that can close. In contrast, in cartilaginous fishes (like the ray, the sea dog, and the shark) the branchial chambers are not shut and can be seen from the outside: these are the "branchial fissures." As water moves into a fish's mouth, it passes through the branchial chambers and washes the gills, which absorb the dissolved oxygen from the water. There are also some fish that can live outside the water for hours. They are considered amphibians and are grouped in two subclasses: the first is the subclass Dipnoi, represented by lungfish that have both lungs and gills; the second is the subclass Periophthalme, or terrestrial fish, in which numerous small blood vessels pass through the mouth, permitting blood to get oxygen from the air. Fish are covered by overlapping scales resembling the tiles of a roof. Sharks do not have scales. Shark skin is rough. This roughness is caused by small tooth-like scales called dermal denticles. They are not visible, but if you touch the skin of a shark, it will feel as rough as sand paper. Fish keep their balance and move through the water by means of fins, which include pectoral, ventral, dorsal (one or more), anal, and caudal (the tail). Except for sharks, fish are almost all oviparous; the female lays her eggs (the spawn) in water, and the male inseminates them. Sharks reproduce by mating, and the female lays already-inseminated spawn or, in some species, it bears already-formed young. Most fish have a swim bladder; this is a more-or-less air-filled organ situated in the body that enables them to float without expending extra energy.

The Batrachians

The batrachians make up a class of animals consisting of approximately 3,000 species, including the frogs, the toads, the salamanders, and the newts. The batrachians are the oldest tetrapod animals (four-legged animals); they emerged in the Carboniferous period (the Paleozoic era) about 350 million years ago. They are amphibians. The name "amphibian" comes from the Greek words amphi, meaning "both," and bios, meaning "life," which means that the animal can live on land as well as in the water. All present-day species live in fresh water or are terrestrial. However, fossils have been found clearly indicating that there were species among the ancestors of present-day amphibians that lived in salt water. The batrachians have a big mouth provided with small hollowed teeth. They have naked skin (without fur, scales, or feathers), and they require humidity; otherwise, the animal dehydrates, dries out, and dies. However, some species adapted pretty well to drought, so that you can meet toad species in the desert, where they spend the day buried in sand. And yet, despite their location, the batrachians need to find some water to lay eggs. The young (called "larvae") resemble little fish; these are the tadpoles. During their aquatic life, the tadpoles transform progressively; the legs appear first (the hind legs first, then the front legs), the fin regresses, and the aquatic breathing by means of gills is replaced little by little by air breathing with lungs. The batrachians' skin is also very important for breathing, because gas exchanges (like oxygen) are made through the skin. Batrachians are cold-blooded animals, just like fish, meaning that they are not able to generate their own body heat. Some batrachians, like frogs and toads, have a sticky protractile tongue, which they use to catch their prey, unfolding it in a way similar to the chameleon.

The Reptiles

The first reptiles emerged on Earth, about 300 million years ago. They were very well adapted to the conditions of that time, and they knew a prosperous evolution. They became very numerous and diversified. Dinosaurs were the most famous reptiles of those times. Nonetheless, the majority of the species from that period disappeared 100 million years ago for one or more reasons that are still not known for sure. (Scientists talk about the cooling of the planet or about a giant asteroid that hit the Earth, and so on). Presently, there are 6,000 species of reptiles on Earth. Except for the snakes and a few lizards that are limbless, all the rest (turtles, lizards, crocodiles) are "tetrapods" (they have four limbs). And yet, the reptiles crawl on the ground rather than walk. This type of locomotion is also suggested by their Latin name reptilis, which comes from repto, meaning "to crawl." As their appendages are very short and positioned laterally (on both sides of the body), the ventral part of their body always touches the ground, even when reptiles move. Reptile skin is covered with tough, horny scales. As the animals grow, the skin becomes too small, and the reptiles shed it: this is called "sloughing." Some reptiles live in water (like marine and fresh water turtles and also some snakes), but the majority are terrestrial. Most of them are oviparous; females always lay eggs in a nest on the ground (the aquatic species come on the ground to lay), but they do not incubate the eggs. Reptiles are cold-blooded animals; their body temperature depends on the temperature of the environment. This is why lizards or snakes can often be seen stretched out and getting "tanned" in the morning sun: this way, they "raise" their body temperature, which decreases during the night.

The Birds

Birds emerged on the Earth by the end of the Jurassic period and the beginning of the Cretaceous (in the Mesozoic era), about 150 million years ago. They come from reptiles. The oldest bird fossil is the Archaeopteryx. This was a very odd animal; it had a lizard-like tail, bird-like feathers, and a beak provided with teeth. Presently, there are about 9,000 species of birds. They are perfectly adapted to aerial life: their body is covered with feathers, and their hind legs have transformed into wings. Almost all of their bones are filled with air. (Such bones are also called "pneumatic" bones.) This characteristic makes the skeleton very light, which is obviously an advantage while flying. Some birds, like ostriches, emus, and kiwis, are flightless, and, instead, they can easily walk or run. Birds were the first warm-blooded animals. This means that their body temperature can remain constant, regardless of the temperature of the external environment. When it's cold outside, they get warm (producing energy), when it's hot they get cool (sweating, for example). All birds are oviparous; they lay and incubate eggs until the young hatch. The egg always has the same basic structure: the yolk and the white (the albuminous mass), both protected in a calcareous shell. The yolk and the white serve as a food supply for the embryo in the process of growth. The embryo needs constant temperature to grow normally. This is why the females incubate their eggs-to assure this necessary constant temperature. Some birds are sedentary (they live all their life in the same place); others migrate (they travel from one place in the world to another). Migrations take place when the seasons change; when winter is near, birds leave for places with milder weather, but they come back in the next warm season. Migration protects them from low temperatures and also enables them to find the food they need to subsist. The insectivorous birds (those that feed on insects), like swallows, for example, need to live where insects are found, and thus are very sensitive to cold weather. As for ducks, they cannot survive if their water source is frozen; therefore, they must migrate during winter to find unfrozen water.

The Mammals

The mammals emerged during the Mesozoic era (in the Jurassic period), 200 million years ago, but they spread and diversified, especially during the Tertiary era (50 million years ago), after the extinction of the large dinosaurs. They conquered all life environments: terrestrial (wolves), underground (moles), aerial (bats), and aquatic (dolphins). Currently, mammals number approximately 5,000 species. They are viviparous animals; their eggs develop partly or entirely in the female's belly, and the young that are born resemble their parents from the very beginning; they do not have to pass through the process of metamorphosis to become adults. In the case of marsupials (like kangaroos and koalas), the newborns must still grow within the mother's ventral pouch. The only exceptions are the representatives of the group called "monotremes"; indeed, primitive mammals, like the echidna and the duckbill, lay eggs just like birds and hatch them in their burrows. All female mammals raise their young on milk by means of mammillae (also called "udders" or "teats"), except for the monotremes, which have mammillary glands that excrete milk. Female mammals always have an even number of mammillae: from one pair (in monkeys and human beings) to ten pairs (in certain insectivores). The number of mammillae can be related to the number of young the female bears. Mammals are warm-blooded animals just like birds, and they keep their body temperature constant. Mammals are the only haired animals. The thickness of their hair varies, and sometimes it can form a thick fur that protects the animal against cold. All mammals breathe with lungs, even the aquatic mammals like the manatee, the whale, or the dolphin. Each mammal species has a specific and varied diet. They can be carnivorous (feeding on meat, like lions or wolves), vegetarians (feeding on plants, like antelopes), piscivorous (fish eating, like dolphins), insectivorous (feeding on insects, like anteaters), carrion feeders (like hyenas), omnivorous (eating a great variety of foods, like pigs and human beings) and even blood-eating (like the vampire bat). Mammals generally have two successive dentitions: milk teeth and final teeth. The final teeth are a usually larger in number than the milk teeth. The final dentition is very important, as it is one of the main criteria of mammal classification. However, certain mammals have few or no teeth at all; they are grouped under the name "edentates," and include the anteater, the armadillo, and the sloth.

Index

Manatee	132	Porcupine Puffer	136	Starfish	130
Marabou	27	Porcupine	41	Starling	267
Margay	81	Prairie Dog	192	Stork	181
Marmot	100	Praying Mantis	85	Striped Ground	
Millipede	265	Puma	97	Squirrel	55
Mole	273	Python	74	Suricate	195
Moloch	52			Swallow	252
Mongoose	48	Rabbit	200	Swan	168
Moose	214	Raccoon	178		
Moray Eel	134	Rat	249	Tapir	76
Mosquito	291	Rattlesnake	46	Tarantula	82
Mouse	289	Ray	138	Tern	150
Musk Ox	123	Razor-Billed Auk	117	Thomson's Gazelle	31
		Red Deer	205	Thrush	211
Northern Gannet	146	Reindeer	111	Tiger	78
Northern Lapwing	182	Rhinoceros	32	Toad	164
		Ring-Tailed Lemur	68	Tortoise	80
Octopus	135	Robin Redbreast	268	Toucan	77
Orangutan	69	Rocky Mountain Goat	94	Trout	187
Oryx	49	Rook	248	Turkey	236
Ostrich	20	Royal Eagle	90	Two-Toed Sloth	71
Otter	177				
Oystercatcher	154	Sable Antelope	34	Viper	198
		Salamander	179		
Pangolin	84	Sandpiper	155	Walrus	122
Parakeet	285	Scarab	272	Warthog	28
Parrot	72	Scorpion	50	Wasp	262
Peacock	250	Sea Horse	131	Whale	126
Pelican	152	Sea Urchin	158	Wild Boar	222
Penguin	118	Seal	114	Wolf	216
Peregrine Falcon	104	Shark	140	WoodLouse	271
Pheasant	237	Sheep	238	Woodpecker	224
Pig	245	Shrimp	159		
Pigeon	263	Skunk	226	Yak	98
Pike	186	Slug	264		
Pink Flamingo	169	Snail	260	Zebra	21
Polar Bear	112	Sparrow	269		
Polar Fox	110	Spiny Lobster	157		
Polecat	212	Springbok	53		
Pony	244	Squirrel	227		

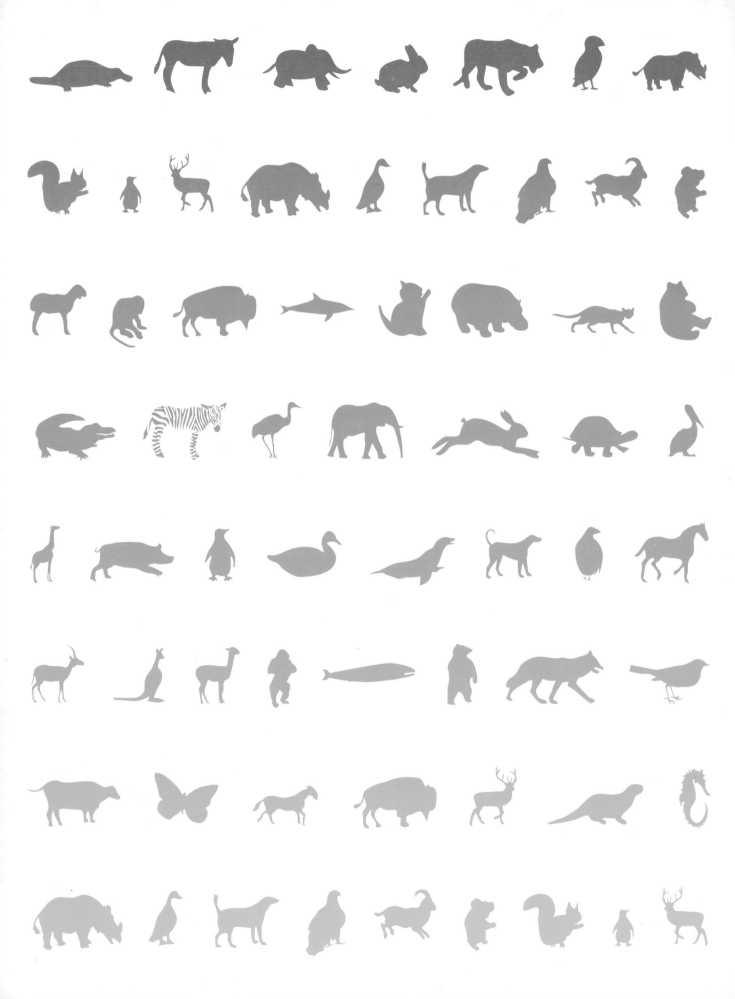